The Balance of Nature

The Balance of Nature

Ecology's
Enduring
Myth

∾

John Kricher

PRINCETON UNIVERSITY PRESS   PRINCETON AND OXFORD

Library of Congress Cataloging-in-Publication Data

Kricher, John C.
  The balance of nature : ecology's enduring myth / John Kricher.
    p. cm.
  Includes bibliographical references and index.
  ISBN 978-0-691-13898-5 (hardcover : alk. paper) 1. Ecology—History.
2. Ecology—Philosophy. 3. Philosophy of nature. I. Title.
  QH540.8K75 2009
  577—dc22   2008044853

British Library Cataloging-in-Publication Data is available

This book is lovingly dedicated to my wife Martha Vaughan, who keeps my nature balanced, and to Sarah Reinstein, my step-daughter. If Sarah is representative of how her generation will steward Earth, there may be reason for optimism.

ॐ Contents

ꙮ Preface

I often enjoy my lunch at a small café named Holly Berry's. It's one of those places where people quickly know your name, at least your first name, where congeniality and food are offered with equal gusto. I was talking to Jack, who most folks would describe as a cook until they taste his food, at which point he'd be better described as a chef. Jack asked me if I was writing a book while on sabbatical and I said that I was, one about the concept of the balance of nature. Jack allowed as to how he had not given a great deal of thought to the balance of nature but, in his line of work, he had thought often about the balance between soup and sandwich. Not just any soup goes well with just any sandwich. And opinions vary. Jack opined that the delicate balance between soup and sandwich is essentially judgmental. In the end it comes down to what tastes good to whoever is having the soup and sandwich. That day I had tomato basil soup and a grilled cheese with onions and tomato. Excellent balance.

The point here is that the concept of balance is, indeed, often in the eye of the beholder. That includes one of the most deep-seated assumptions about balance, that there is such a thing as the balance of nature. Such a realization is nontrivial.

This is a book about ecology but it is not the usual kind of book about ecology. There are several themes that run through the narrative. One is an account of what ecology is today and something of how it got to be that way. I describe how ecology has emerged and matured from observation and description of patterns in nature to become a more predictive and pragmatic science, albeit one that remains empirical and, as I will argue, essentially without paradigms. A second theme is how ecology has only recently shed some of its most burdensome philosophical baggage, the concept that there is a natural "balance of nature." And a third theme is why it is incumbent on citizens to seek some understanding of ecology, particularly as it contributes to understanding climate change, various conservation issues, and the complex matter of how to treat nature's biodiversity.

Ecology is a branch of evolutionary biology. This is because any form of biology is, in reality, a branch of evolutionary biology. Failure to have some understanding of how evolution works makes understanding biology, and, today, understanding what is happening in the world, very difficult. Ecologists study how Earth's ecosystems function. The relevancy of this information goes well beyond the ivory towers of academe and should be part of every educated citizen's understanding of his or her world.

This book is designed for a broad audience, particularly readers who intellectually thrive by seeing connections among various disciplines such as philosophy, history, and the sciences. Ecology has deep philosophical roots, especially in the study of natural history, and has enjoyed a vigorous growth in the previous century. Hopefully it will distinguish itself in the present century.

I have studied and taught ecology for nearly four decades and I am opinionated about it. That reality will be evident throughout this book. I tell my students that questions dealing with ecology will accompany their lives throughout the present century. The Earth's collective biodiversity is experiencing its sixth major extinction event since multicellular life first evolved. It is also passing through a period of rapid climate change, a change that is likely forced by anthropogenic actions. The natural world of the twenty-first century may bear little resemblance to that of recent centuries. We have entered the Anthropogenic era in which human influence is pervasive in all ecosystems.

My wish is that this book will help the reader to understand the connections between evolution, ecology, and other areas of human thought and realize that "nothing endures but change." However, some changes are better than others, and the great virtue of being humans is that, at least in theory, we have a choice. Our destinies are largely under our control if we seize that control. Ecology is no longer the arcane study of natural history. Ecology, in the twenty-first century, may be the key to human destiny in the twenty-second century and beyond.

The Balance of Nature

∾ 1

Why It Matters

What is this Balance of Nature that Ecologists Talk About?
—Stuart L. Pimm, *The Balance of Nature*[1]

That there is a balance of nature is one of the most deep-seated assumptions about the natural world, the world we know on planet Earth. For as long as we humans have had the ability to think seriously about our world we have attempted to find order in chaos. The world is vast and surely appeared vaster when our collective knowledge was far less than it is today. Humans living, say, 10,000 years ago, at the dawn of agriculture, must have perceived nature as impossibly complex, perhaps beautiful, very mysterious, and surely fairly scary. These perceptions have changed to various degrees. Today *Homo sapiens* has emerged as the dominant species on the planet, as measured by its collective effects on Earth's ecosystems. No single species in Earth's history has caused more changes on the planet than what we are doing today. We need to understand and act on this reality. But why? Begin by allowing me to take you on a journey beyond Earth, through a bit of space and time, and you'll soon see "why it matters."

We live in the Stelliferous era, the time of the stars. There was a previous time when there were no stars, and there will be a time in the far distant future when there will be only cold and dark remnants of stars, when absolutely no form of life will exist anywhere in the universe. All traces of human existence or any other forms of life will presumably have long since disappeared from the cosmos. The universe, our universe, will be dead. The very early history of the universe in which we reside was one of short-lived symmetry and order, lasting but the tiniest fraction of a second, until small asymmetries made possible the eventual formation of elements, stars, and galaxies in a universe fated to expand forever. The universe, and all in it, has essentially been asymmetrical and evolving ever since.[2]

Much of the universe is a violent place. Stars periodically explode, becoming gargantuan supernovas, then collapse, shedding their outer gases to end their stellar existences as cold, dark dwarfs, some of which, the pulsars, spin around at a dizzying pace, curious space beacons in the aftermath of trauma. Immense black holes lurk menacingly in the center of galaxies, astronomical quicksand sucking up the stellar offspring of the big bang. Our own Sun is no less than a consolidation of billions of constantly exploding hydrogen bombs, a thermonuclear furnace, continuously engaged in the most violent reaction known, the result of which keeps us warm, gives green plants their most vital ingredient, and tans our skins. Yes, the universe is violent and basically unpleasant when you get too close. It's pretty hostile outside of the spacecraft. Thinking about the stunning forces that govern, indeed define, the universe can make humanity and life in general seem very frail.

Stars form, stars shine, and stars go dark, their nuclear fuel fully consumed. Such a fate eventually will befall our star, the Sun. These billions upon billions of huge, gassy gravitational concretions of concentrated thermonuclear energy just come and go. All that is required is time. Lots of time. And there has been lots of time. The universe is estimated to be 13.7 billion years old,[3] roughly two-thirds again as old as our Sun and its solar system. And the universe will become much, much older.

In July 1994, the planet Jupiter was repeatedly struck by pieces of Comet Shoemaker-Levy, whose path had been altered by the huge gravitational field of the planetary giant, itself a stillborn star.[4] Jupiter literally pulled the comet from space, shredding it in the process, and pieces of the comet left obvious impact marks across the face of the planet. If that happened to Earth (as it has in the past), it would be bad. Are we safe? No, we aren't. It's a matter of probabilities. We've been hit before and will likely be hit again (chapter 10). Even in our local solar neighborhood among our sibling planets, the threat of violence is lurking. Maybe there's good reason to be afraid of the dark.

But on the other hand, on a warm summer night when it really is dark and that magnificent and vast assemblage of stars we call the Milky Way traces its winding course across the sky, we humans perceive the universe around us as anything but violent. Seeing dust-sized meteors streaking through the atmosphere leaving a momentary trace of firelike light does not inspire thoughts of imminent doom from asteroid impact. There is, more often than not, a sense of profound tranquility and serenity imparted to one's psyche when lying on one's back in the cool, damp grass and staring skyward at the slowly revolving panorama of thousands of points of light so far above, so far away. From our earthly perspective, the universe can seem ever so peaceful, constant, predictable, and essentially inviting. The phases of the Moon change, but do so in a most orderly, predictable fashion. The Sun never disappoints, always faithfully rising in the east, setting in the west, never the other way around. The constellations seem unchanging (though they are not—again, it's only a matter of time), and the planets predictably trace their respective paths around the Sun, moving through the Zodiac from constellation to constellation (regrettably keeping astrologers in business).

The universe is exquisite, never minding the copious unbridled forces underpinning and sustaining the apparent beauty. It's really no wonder that heaven, as it is envisioned by those who believe there is such a place, should be located somewhere in that cornucopia of glittering stars. The violent universe deceives us, seeming to welcome us, a falsely serene place, its parts working in the illusion of harmony, the so-called "music of the spheres."

The point here is that things, including natural things, are not always as they seem. Nor are they necessarily as we might wish them to be. They just are. One's perspective on the universe can be highly quixotic, a perception that satisfies, that makes us happy when thinking about the heavens above, even if not very accurate. Or, one might envision the universe exactly as astronomers say it is, with all the accompanying violent reactions that have defined stellar birth and death from the big bang onward. In either case, one can be fascinated and mentally fulfilled just thinking about what's out there.

As far as the universe is concerned, it really doesn't matter what we think. Think anything you want. Whatever the universe is or isn't, there isn't anything you or I can do about it. We can't blow up Saturn, pollute the Sun, or cause the extinction of any of the stars. The clusters of galaxies will continue to fly apart from one another as spacetime expands, whether we approve or not. We have no power whatsoever to influence events occurring tens, hundreds, millions, or billions of light-years away from our own planet. We humans are utterly trivial in our collective influence on the workings of the universe we inhabit. A fly has more effect on the atmosphere of the Earth than we collectively have on the universe.

But, on the contrary, it does matter what we believe about the workings of our own planet, endowed as it is with a myriad of living systems we call organisms, including in excess of six billion human beings. We can and do affect the ecology of the Earth in innumerable and profound ways. If what we do is wrong, it will have consequences and costs. It already has. What we do is obviously largely determined by what we believe about the structures and workings of the systems we affect, so, to say it again, what we believe about Earth, ecology, nature, and our own biology and evolution, matters.

We are beginning what is called the twenty-first century. It isn't really. There have actually been 45 million centuries in the history of this planet, but we anthropocentric humans pretentiously identify only those that began with the birth of Christ plus a few hundred that preceded that particular historic event. In reality, for every year in which *Homo sapiens* has inhabited the planet (assuming approximately 100,000 years as "modern humans"), there have been roughly 45,000 when it was absent. Nonetheless, we are here now and the twentieth century was perhaps most notable for the degree to which one species, the human species, has influenced the Earth's ecology. Never in the 4.5 billion-year history of Earth has but one species had such an inordinate influence on all others, and in such a short span of time. That influence grows every second.

It is the objective of these essays to examine what is known about some important aspects of how life on Earth functions. One theme will weave throughout the essays, connecting them as the message of the book: there really is no such thing as a "balance of nature." Nor is there purpose to nature. Nature, like the vast universe of which it is but an infinitesimal part, just is.

That our perception of nature may be erroneous is not a trivial point. It is, indeed, very critical to know how nature works. It does matter. Our welfare as well as that of many, and perhaps most other nonhuman life forms, ultimately depends on judgments we make based upon what we know about the workings of the biosphere, that thin layer of life that coats Earth's surface.

I don't believe in Santa Claus, though once I did. I think it does no harm to tell a child a myth about a red-suited, white-bearded, elderly philanthropist who likes hot cocoa. Realizing that the myth is, indeed, a myth is not really very traumatic, at least it wasn't for me. I hold no grudge against my parents for helping promulgate an enchanting falsehood. Quite the contrary, those were good Christmases, leaving me with the best of childhood memories. During the time when I believed in Santa Claus I took a nasty fall and a rusty nail penetrated deeply into my scalp. I soon developed a raging case of septicemia, and might have died. However, my parents saw to it that I quickly got to a doctor and was treated with massive doses of penicillin. After a few bad days, I recovered. I'm really glad my parents believed in medicine, not spiritual healing or something like it. My life was saved by a chemical evolved by a fungus in response to the collective competitive pressures exerted over eons by bacteria, the chief competitors of fungi. I owe my life to an evolutionary by-product of interspecific competition, from a species of mold, the properties of which, incidentally, were discovered mostly by accident. Millions of us owe the same debt. When it comes to life support systems, it won't do to create myths.

Knowledge is not easily acquired. It is far simpler to believe than to discover. To give but one example, Ancient Egyptian mummy preparators, who were otherwise pretty good at what they did, routinely extracted the brain through the nostrils and discarded it, utterly ignorant of its function and profound importance.[5] Other organs were carefully saved and prepared, to accompany the pharaoh on the journey into the afterlife. Poor pharaoh. Eyes, ears, nose, heart, liver, lungs, body, but brainless. Not much of an afterlife. Imagine the deceased royalty of ancient Egypt all mingling about in the Great Beyond with nothing to say to each other but "Duh."

It required centuries of medical study and experiment to learn that the heart is not the center of the soul but is instead a sophisticated, coordinated blood pump. How the brain works is still far from fully understood. But this much we do know: we think, we feel, we love, we hurt, we hunger, and we believe with our brain. The ancient Egyptians were wrong. Their view of human physiology was flawed. Some contemporary cultural relativists, abounding as they do in the halls of academe, might argue that the ancient Egyptian view was "equally valid" to the modern view, and should be "celebrated." Celebrate it all you want, it's still wrong.

And there is something else in the example to note. Science is a way of knowing. It is actually possible to get the right answer, though many wrong ones may crop up along the way. Since scientific truths must be discovered, and since many, probably most, are far from intuitively obvious, wrong answers are inevitable. The path to the truth is sinuous, not easily navigated. One reason for such difficulties is that scientifically gained knowledge is often nonintuitive or even counterintuitive. In the vernacular of some college students, "science is hard." However, with a reasonably open mind and persistence, right answers and understanding are achievable.

From the time of early human civilization, most notably the intellectual contributions of the ancient Greeks, humans have envisioned life on Earth as having both balance and purpose. Such a notion was philosophically satisfying, immensely so, perhaps even essential for the psyche of those toga-clad early thinkers. It was supported, albeit at a superficial level, by lots of observational evidence. There are many people today who harbor similar beliefs. Creationists, now reinvented as "students of intelligent design," continue their efforts to make science subservient to religious dogma, as they try harder and harder to philosophically pound a very square peg into an awfully round hole. As a second example, conservationists who believe fully in evolution, including human evolution, worry about upsetting the balance of nature, causing irreparable harm to Earth's life support systems. Most people in the United States are regrettably ignorant about what is known in ecology as it relates to evolutionary biology, and how this information, this knowledge, these facts, should affect decision making about environmental issues.

It will be my task to convince you that life on Earth has neither innate balance, nor purpose, at least in the meanings usually associated with those words. It is not my intention to demean human existence or that of any other species. Quite the contrary, I wish to focus on the importance of understanding how life functions evolutionarily and ecologically so that our species can assume a more realistic and ultimately more responsible role in its task of stewardship of the planet.

Philosophers have noted that scientific truths should not, in themselves, lead to prescriptive ethics. The so-called "naturalistic fallacy" asserts that one should not assume that what is, is what ought to be.[6] The naturalistic fallacy was conceived to separate science, especially evolutionary biology, from philosophy, especially ethics. However, in the latter part of the twentieth century, the two disciplines came increasingly closer. Some philosophers now refute much of the naturalistic fallacy. Ecosystem restoration and management, based on the science of ecology, is applied to moral decisions about whether or not we ought to try and preserve endangered species. Studies of animal behavior and molecular genetics that indicate a profound Darwinian link between humans and apes raise significant moral questions about whether sentient or even partially sentient nonhumans should endure medical experimentation.

In my view the time has come to free ourselves from some notions that originated almost as early as civilization itself, notions that have, in my opinion, become more of a hindrance than a help. We still carry too much philosophical baggage. The time has come to leave some of it behind.

∾ 2

Of What Purpose Are Mosquitoes?

Only once have I been accused of being a baby killer. To be absolutely accurate, I was accused of being an accomplice. The actual murderer was an anonymous mosquito. Oh, and the murder had not yet occurred, but surely would unless this particular small New England town dowsed itself with insecticide. Spraying would hopefully annihilate the summer hordes of mosquitoes, a few of which could possibly be carriers of the virus of eastern equine encephalitis, a tiny chunk of renegade nucleic acid that unfortunately can sometimes kill not only babies but other people as well.

As a scientist and academician with some experience studying wetland ecology, I was asked to comment at a town meeting on the effect that spraying for mosquitoes might have on the local ecology. It is important to note here that the virus in question had not yet been shown to be present in that summer's mosquito population, the factor that led to my qualified recommendation against wholesale spraying. This advice was anything but satisfactory to some of the local citizens, many of whom seemed agitated and one of whom rose among the tumultuous crowd and stridently asked if I wanted to kill her baby. I guess I was perceived by the majority as a foolish intellectual, remote from the concerns of everyday people, an elitist defender of potentially lethal vermin. This is a difficult and uncomfortable position in which to find oneself.

After the meeting a friend offered solace. She said she was certain that mosquitoes have some purpose in the great balance of nature (a balance that spraying would perhaps upset), but that I just hadn't been able to explain what that particular purpose happened to be.

That was a kind of sympathy I didn't want. I don't believe mosquitoes have any purpose whatsoever. Nor do I believe that any other form of life has purpose, including humans. The closest thing to purpose that applies to life forms is reproduction, but reproduction is not really purpose; it is, like all else in physiology, function, a part of metabolism and the life cycle. The urge to reproduce, as well as the anatomy and physiology enabling reproduction, is coded in the genes. If mosquitoes do not make more mosquitoes there will be no mosquitoes. Mosquitoes know this, so to speak, because each and every one of them is a faithful servant of its genes, which are molecules of DNA (deoxyribonucleic acid), a compound with the astonishing property of accurate (if not always perfect) self-replication. DNA made the mosquito, including its brain, sense organs, appetite, and urge to procreate. A mosquito's actions are literal manifestations of an elaborate set of lifetime instructions encoded in its genes. The insect faithfully and blindly obeys its DNA's directives during its brief tenure on the planet. Mosquitoes, like every other organism, are only immortal as genes, not as individuals.[1]

It is not appropriate to define reproduction as having purpose, any more than it is to define breathing as purpose. It is no more accurate to say you are put on Earth to reproduce than to say you are here to inhale and exhale air. Life is. Life reproduces. Life respires.

The distinction between "purpose" and "function" is nontrivial in the context of science. In driving an automobile it may be acceptable to say the *purpose* of the pistons is to provide kinetic energy to power the drive shaft, just as it is equally fine to say that the *function* of the pistons is to provide kinetic energy to power the drive shaft. Automotive engineers intelligently design cars. Thus purpose and function, in this case, are essentially synonymous. But organisms are not cars. They evolved, thus the word "purpose" implies something that is not accurate, at least not to evolutionary biologists. Mosquitoes were not designed with any purpose in mind because they were not designed in the first place. They function as mosquitoes because they evolved through the blind and mechanistic process of natural selection.

Humans got here through the exact same biological process that made mosquitoes, water buffaloes, woolly mammoths, canaries, and every other life form that's now and has ever been. Asserting our species' biological lack of purpose does not, as some might be quick to suggest, automatically mean one believes that human existence is meaningless. Quite the contrary. We have a uniqueness, evolved somewhere between 500 and 1,000 centuries ago, that profoundly sets us apart from the remainder of Earth's biota including our less intellectually gifted immediate ancestors among hominins.[2] We are truly sentient. It is hard to see how anything could be more meaningful.

My objection to wholesale mosquito spraying was pragmatic. Mosquitoes, like all other living things, are subject to natural selection, discovered and described in the nineteenth century by Charles Darwin and Alfred Russel Wallace (chapter 5). Natural selection is largely responsible for why living things look, act, sound, and smell as they do, or, in the case of extinct forms, as they did. DNA has the property that it occasionally mutates. Most times, mutations are harmful or at least do no good. Sometimes, however, a mutation may enhance survival in certain situations. The more mosquitoes are exposed to pesticide, the more the pesticide selects for the most pesticide-resistant mosquitoes among the population. Those few with genes that enable them to tolerate the esoteric insecticide are, of course, not affected by it (or less affected by it). These genetically resistant individuals remain healthy, survive the spraying, and outbreed the nonresistant mosquitoes (which are sick, dying, or dead, in each case a significant impediment to breeding), and thus the next generation contains a far higher proportion of resistant individuals.[3] Oh sure, you may kill 90% or even 99% of the population, but the survivors, though initially far fewer in number, will provide the entire next generation, a generation composed of far more resistant animals. Soon the population rebounds and spraying (at least of that particular insecticide at that particular dose) becomes useless. If the insect is vectoring some pathogen, such as eastern equine encephalitis, and the majority of the insects are pretty much resistant to insecticide, it's time to worry. There are some areas where mosquitoes are resistant to virtually all commonly used pesticides, even when they're applied at very high dosage.

The same argument applies to the use of antibiotics, which are now administered far more conservatively than in past years. Antibiotic exposure selects for resistant bacteria, just as insecticide exposure selects for resistant insects. Antibiotic-resistant bacteria can kill you and there is not much you can do to stop them. And more and more resistant strains of bacteria are evolving.

But back to mosquitoes. They do carry eastern equine encephalitis (often simply referred to as "triple-E"). They also carry, depending upon geographic location, various forms of malaria, yellow fever, West Nile virus, dengue, and a number of other unpleasant maladies that, taken together, have likely killed far more millions of people than all the horrendous human conflicts throughout history. For example, the World Health Organization reports that about 500 million cases of malaria occur annually (with about a million fatalities) and 2.37 billion people in 87 countries risk contracting malaria. Mosquitoes can indeed be dangerous.

Since no cases of eastern equine encephalitis had yet occurred in the summer when I had my traumatic confrontation at the town meeting, I suggested that it was ill advised to take the risk of making a genetically stronger mosquito when there was, as yet, no demonstration of a real problem. I was quite literally shouted down. They sprayed.

Mosquitoes are insects and insects are the most diverse and abundant of the many groups of animals. Insects are generally small in body size (the largest one known, a dragonfly-like creature that lived about 350 million years ago, had a wingspan of about a meter). Small size is an adaptation of sorts, since being little means there is plenty of room for insects on the planet, and they have certainly availed themselves of the space. They can be amazingly specialized, some living within leaf veins, some under tree bark, some manufacturing honey, or constructing elaborate nesting chambers, or laying eggs on fresh wounds, or burying dung, or pollinating roses. Mosquitoes are capable of living in all manner of habitats: swamps, tundra, grasslands, deserts, bromeliad leaves in tropical rainforests, tree cavities, cans and pots, old discarded tires, and basically any type of puddle, including bovine and equine hoof prints. All mosquitoes, and there are about 3,500 species, are blood feeders. Or at least the females are; males confine their sucking activities to plant juices. With apologies to the illustrious count from Transylvania, blood feeding may seem odd and nasty from our point of view, but it makes adaptive sense that female mosquitoes have an appetite for blood, since they require prodigious amounts of protein to make healthy eggs, and blood is ever so full of protein. Blood sucking is thus adaptive in helping female mosquitoes make robust eggs that grow into healthy mosquito larvae. Adaptive or not, it is still not an endearing trait.

Mosquitoes, even when not transporting fatal diseases, are nonetheless generally irritating. Most of us know this from personal experience. I was in Bali a few summers ago at a small town called Ubud. There were no screens in the windows and my room abutted a rice paddy. It was night and I was attempting to get to sleep in this tropical paradise when I detected the disconcerting and persistent whine of a mosquito hovering somewhere in the vicinity of my left ear. Soon a good few of its colleagues joined it, flying sortie after sortie, as I waved my arms in the dark. I kind of envisioned myself as King Kong atop the Empire State Building swatting at those pesky little airplanes. I had forgotten to light the mosquito coil that would have provided a constant dose of pyrethrum, something that hungry lady mosquitoes apparently find suitably repulsive. Given that *Plasmodium falciparum*, one of the most lethal and drug-resistant species of protozoa that cause malaria, was known to occur on Bali, I thought it might be wise to get out of bed and light the coil, so I did. Once that was done, the mosquitoes left me pretty much alone for the evening.

So why are mosquitoes? Wrong question. Asking why puts an assumption of purpose in the question. But what's wrong with that? Here's what's wrong with it. Assuming that I have purpose, you have purpose, life in general has purpose, then each of the many individual forms of life ought to have purpose, and thus mosquitoes, all 3,500 species of them, ought to have purpose. So why are they here? Are mosquitoes here to fulfill some important, perhaps crucial, role in the great balance that is nature? If mosquitoes were to suddenly disappear from Earth, would the dominoes of life begin to fall?

Such a question, like much of science, is more complex than it first appears. As I will describe in later chapters, species vary in the degree to which they influence the stability of the ecosystem of which they are a part. Are mosquitoes "load-bearing" or "keystone" species, whose disappearance would profoundly alter the ecosystem, or not? It is essential to know these sorts of answers if humans are to act as prudent stewards of the planet.

Creationism, at least the variety of it that fundamentalist Christians adhere to, asserts as fact that God literally created all forms of life as is. But if this were true, you sort of have to wonder why God made these little bloodsuckers, and so many of them at that. Then there are the horseflies, ticks, and leeches, and those are just some of the parasites that live on the outside of us. Look inside the corpus and there reside the botflies, leishmanias, plasmodiums, shistosomes, trypanosomes, tapeworms, flukes, and roundworms. Apparently God enjoys making parasites.

And that's not all. God's apparent fixation with making insects has not gone unnoticed. The British evolutionist J.B.S. Haldane was once asked by a clergyman what Haldane's study of evolution had taught him about the will of the Creator. Haldane is said to have replied, "He has an inordinate fondness for beetles." Fondness indeed.[4] Of the 1,032,000 described animal species currently resident on the planet, nearly three of every four of them, some 751,000, are insects, and of those, over 300,000 are in the order Coleoptera, the beetles. Just considering dung beetles alone, there are, at last count, about 2,400 species—and these are insects whose life cycle is based on finding and burying feces! Thus, if God created all animal species as is, He made one out of every three species a beetle (and these numbers are likely to be significant underestimates of the actual global species diversity, which as yet is not precisely known for insects, and may never be known with certainty). Beetle species outnumber higher plants, of which there are about 248,000 species, as well as chordates, with a meager 43,000 species, over half of which are fish. It is probably obvious by now that I am not a believer in the tenets of creationism, thus I am relieved of the burden of explaining why God is so fixated on six legs, an antenna, and large compound eyes.

Most biologists believe that mosquitoes evolved in an ongoing process largely accounted for by the one universal law of life, *the* biological paradigm, natural selection. Does evolution by natural selection imply or suggest purpose? Sloppy science writers seem to think so. It is not at all uncommon to read about how bird wings are *designed* for flight, that their *purpose* is to provide thrust and lift, even when the author is actually explaining natural selection, using bird wings as an example!

Some might say that evolution has some innate direction, perhaps part of a great plan of life where all living things evolve together, each with its role to fulfill. Don't believe it. Natural selection is nothing more than a statistical game of genetic survival, what has been termed the "ultimate existential game." You have no choice but to play (though, uniquely for humans, you do have a choice) and you can never really win, only earn the right to go on playing (i.e., the best you can hope for is to survive long enough to reproduce and raise your kids to reproductive age). By this reasoning, mosquitoes are seen as temporary repositories for combinations of genes that have managed to both manufacture and use the diminutive vampires to wander their way through time, beginning the trek many millions of years ago, when dinosaurs still occupied most of the world's terrestrial ecosystems (and thus becoming the ideal fossil vehicle for snagging dinosaur DNA to eventually populate *Jurassic Park*).

Of the many cliches that pepper our minds, one that has a large grain of truth is that nature abhors a vacuum. In other words, no potential habitat remains unoccupied for very long, though some habitats are certainly far easier to colonize than others. Many life forms are adapted to colonize newly opened environments, soon transforming the lifeless place into a thriving ecosystem teeming with various bacteria, protozoa, fungi, plants, and animals. In most cases (the deep ocean heat vents being one exception), these myriad life forms ultimately survive by obtaining energy captured from the Sun in a process termed photosynthesis. But only a few life forms, certain bacteria and protozoa for instance, as well as the vast majority of green plants, have the physiological ability to capture solar radiation and combine it with atoms to make high-energy compounds such as glucose. We can't lie out in the sun and get fat. Corn does. Thus, we eat corn to get the sun's energy (or we eat beef or pork or chicken that has been fed corn). So not only does life abound, not only does biodiversity fill ecological vacuums, but life forms soon come to intimately depend on other life forms for food.

The motion picture *Popeye*, not very memorable for most of us, does contain a delightful song with the biologically accurate title, "Everything Is Food, Food, Food."[5] How true. A living squirrel in the woods is fed upon by mosquitoes, black flies, ticks, mites, and hosts of other ectoparasites. Its skin may be nourishing several kinds of fungi. Inside, the squirrel may be sharing its digested acorns with any number of intestinal worms and protozoa. Its blood may support a community of still other protozoan parasites. Then along comes a red fox with an appetite, and, in a quick ambush and sharp bite to the neck, it efficiently turns out the squirrel's lights and chows down on the still warm bushy-tailed corpse. The satiated fox leaves the body, which soon hosts fly eggs and, shortly thereafter, a thriving population of maggots. Bacteria and fungi also quickly colonize the disappearing squirrel. Its remaining muscles, sinews, skin, and guts become a stew for microbes. Finally there is only a skeleton, and the bones have so little organic matter that nothing much has evolved to eat them, so they tend to stay around. In nature, a little squirrel can go a long way, calorically speaking. Energy from the Sun made acorns that were food for the squirrel. That solar energy moved through dozens, even hundreds, of creatures that utilized the organic matter in the body mass that was a squirrel. Such interdependency among myriad creatures is impressive, but it should not be misinterpreted as some form of balance of nature.

Food provision is far from the only kind of interaction in nature. For example, some orchids have flowers that bear a striking resemblance to the backsides of bees, a characteristic that attracts male bees whose DNA is telling them to make every effort to get their DNA into the posteriors of female bees. The male bees poking into the orchids inadvertently pick up sticky pollen from an orchid flower and, because they are slow learners, fly off only to drop the pollen on yet another orchid flower that has fooled them. Without the bees, the orchid doesn't reproduce.[6] It depends, as do thousands of flowering plants, on an animal vector to move sperm-containing pollen from one plant to another, cross-pollinating the plants. Such nearly countless examples of interdependence between species are the stuff that makes natural history so fascinating. It also gives rise to the not so subtle notion that because things are so clearly interdependent, they must exist in a sort of balance, a balance of nature.

Historically, the notion of a balance of nature is part observational, part metaphysical, and not scientific in any way. It is an example of an ancient belief system called teleology, the notion that what we call nature has a predetermined destiny associated with its component parts, and that these parts, mosquitoes included, all fit together into an integrated, well-ordered system that was created by design. Such a belief in the harmony of nature requires purpose, a purpose presumably imposed by the goodness and profound wisdom of a deity (or deities). Such a view of how nature functions dominated human thought for millennia. For many, likely most, it remains a worldview today.

In the more modern sense, the continued perception that nature is structured in some sort of balance results from what ecologists call "scale effect."[7] Ecologists, as we shall see, were slow to come to this realization, and thus the balance of nature idea was permitted to move unscathed from its teleological roots to become assimilated into materialistically based science. If you look at nature on a very small scale, like a plant and its pollinators, things can appear quite balanced indeed. But this balance is illusory, the result of natural selection making organisms act opportunistically in their own self-interest, such that their fates become interlocked.

Organisms have always been resources for other organisms ever since life began. The most intricate mutualisms in nature can usually be just as satisfactorily explained as cases of mutual parasitism. It is no more accurate to say that a bee is cooperating with an orchid to spread its pollen than to say that a moose is cooperating with a wolf pack by allowing itself to be torn to shreds so wolves can eat.

You can also observe nature on a larger scale and erroneously conclude that there is a real balance operating. For many years ecologists have believed that ecosystems such as forests pass through a series of successional "stages" eventually to attain what is called a "climax" condition, where the biodiversity of the forest is in a kind of stable, long-lasting equilibrium. This notion, which I will explain in more detail in chapter 6, is largely discredited today, and forests (as well as other kinds of ecosystems) are seen as dynamic and changing. However, quite possibly because trees have significantly longer lives than humans do, the initial perception of forests was that of balance and stability.

Now, back to mosquitoes yet again. Mosquitoes are food for many birds such as purple martins (*Progne subis*). These sleek swallows, as well as other bird species, consume immense numbers of mosquitoes. Numerous fish species gorge on mosquito larvae. Without mosquitoes, and many of them, what would become of these creatures? They would either switch to some other food sources, or move elsewhere in search of food, or, least likely of the possible outcomes, just go extinct. The loss of mosquitoes would certainly have measurable impacts on some species. Diversity and abundance patterns would change. Some species might be strongly affected, but some, perhaps many, would not be affected at all.

Those species with the strongest dependence on mosquitoes as a food source would also be affected by insecticide application, as they would consume the insects that contain the toxin and thus begin to move it throughout the food web. In this manner, the concentration of insecticide increases dramatically as it passes from small consumers to top carnivores. It is well known, for instance, that such species as peregrine falcons (*Falco peregrinus*), brown pelicans (*Pelecanus occidentalis*), and ospreys (*Pandion haliaetus*) all suffered precipitous declines in population before DDT and other persistent chlorinated hydrocarbons (used to spray for mosquitoes as well as other insects) were banned.

Given that ospreys, which never eat mosquitoes, are nonetheless dependent on them (from mosquito larvae to tiny fish to medium fish to big fish to osprey to osprey eggs), should we protect mosquitoes in order to ensure the future for ospreys? Is it the purpose of mosquitoes to ultimately provide calories to large birds of prey? Suppose the mosquitoes in question are vectors of the malaria protozoan? Protecting these mosquitoes would not only ensure food for ospreys but also conserve another species as well, the plasmodium protozoan that is malaria. But plasmodia feed, among other things, on the red blood corpuscles of human beings, causing a debilitating disease in the process. Do ospreys have a right to pesticide-free fish if it means malaria-infected humans? Do mosquitoes have a right to thrive because many different organisms depend on them? Is it the purpose of mosquitoes to form a base for the osprey food chain or to provide shelter for plasmodia? For that matter, do plasmodia have an ethical right to access mosquitoes and humans, both of which are necessary for them to complete their life cycle?

Two ideas are deliberately intertwined in the preceding paragraphs. One is the notion of balance of nature in the scientific, ecological sense, the notion that there does exist an actual, measurable, "normal" state in nature in which populations are sufficiently interdependent as to be accurately described as being in a state of balance. Disturbing this balance would be expected to create a cascade of effects, mostly negative, because implied in the notion of balance is that the balance will be optimal. This is the analogy with a machine (nature), whose component parts (organisms) are arranged in such a way as to make the machine function properly. Disturbing this arrangement will, in all likelihood, reduce the efficiency or power of the machine, or even render the machine inoperable.

The second notion is that organisms must have intrinsic value, presumably because they are all part of nature's skillfully crafted and complex machine. Your heart has great value to your body. No one would argue that point. Does a mosquito population have similar value to a marsh? If the organisms in a marsh were as complexly interdependent as the organs that make up your body, the answer would seem to be yes. But maybe a mosquito population is more like a toe than a heart. Maybe a marsh could get along without mosquitoes, just as you could function without one of your toes. The issue then becomes one of environmental ethics. What value, if any, should be placed on various organisms? Is this value dependent on the importance of the creature to its environment as ecologically determined, on the assumption that it must be part of a grand balance, on its appeal to us humans, on its threat to us humans, or on some other set of parameters?

People routinely make value judgments about nonhumans. Someone who would think nothing of flushing a goldfish down the toilet would be enraged and repulsed to learn that his neighbor just flushed a kitten down the toilet. A hunter pausing to admire a bright red male cardinal singing from a shrub would think it quite reasonable to quickly turn and shoot the woodcock that flew out from the base of the shrub. What are the criteria for making such value judgments? Do ospreys have more of a right to exist than anopheles mosquitoes? If so, why? If not, why? Such questions derive from how the human species, the sentiently unique species among all life forms that have ever inhabited the planet, regards what it has come to call nature.

What is most remarkable about the balance of nature idea is its longevity. It has always been a fuzzy, poorly defined idea that nonetheless has had great heuristic appeal throughout the ages because it seems so self-evident. Always more of an assumption than a demonstrated scientific principle, the balance of nature remains in the minds of many people an uncritical paradigm for how they view nature. Once this paradigm is discarded, what consequences arise? Just what is nature anyway?

✎ 3

Creating Paradigms

The "balance of nature" is a paradigm, a venerable and little-questioned belief about how nature is organized. Almost anyone will tell you they think there is some kind of "balance" in nature and that humans tend to upset that balance. Numerous websites are devoted to it, and the history of the concept has been well documented.[1] Humans create paradigms for a number of obvious reasons. We wish to make sense of our world as well as the universe of which it is part, but in doing so, we wish to simplify and unify information that, at first glance, appears to be hopelessly complex and disparate. We also wish to feel empowered, to have the sense that we really know about something of major significance to us.

A paradigm is an all-encompassing idea, a model providing a way of looking at the world such that an array of diverse observations is united under one umbrella of belief, and a series of related questions are thus answered. Paradigms provide broad understanding, a certain "comfort level," the psychological satisfaction associated with a mystery solved. What is important here, and perhaps surprising at first glance, is that a paradigm need not have much to do with reality. It does not have to be factual. It just needs to be satisfying to those whom it serves. For example, all creation myths, including the Judeo-Christian story of Adam and Eve in the Garden of Eden, are certainly paradigms, at least to those who subscribe to the particular faith that generated the myth. Creation myths "explain" perhaps the biggest question ever posed by humans, how did we get here? Can the many varied creation myths of the world's diverse cultures all be literally correct? Are any literally correct? It is, of course, most likely that none is accurate, but each has served its particular constituency for venerable lengths of time. Historically, paradigms have been assembled in various ways, most deduced from historical observation followed by extrapolation, usually with a large measure of intuition and poetic license added by "great thinkers." Over the course of the past twenty-five centuries or so, a uniquely distinct kind of paradigm development has gradually emerged. We call it science.

Scientific paradigms differ from other belief systems in that they not only answer but also, more importantly, generate questions, predictions that are testable in such a way that the paradigm may be validated, revised, strengthened, or discarded. For example, quantum theory suggested the existence of the tau neutrino, a virtually imaginary and bizarre particle that, after years of searching, was actually discovered, a compelling confirmation of yet another component of quantum theory. Thus scientific paradigms are usually associated with scientific "revolutions," where one paradigm sweeps away another.[2]

Scientific paradigms are belief systems based not upon faith in any absolute sense but rather upon a body of data subjected to inductive analysis, and constantly subject to creative skeptical inquiry and challenge. As the physicist Richard Feynman has written, "Learn from science that you *must* doubt the experts."[3] Feynman was describing a core principle of scientific paradigms, as elucidated by the philosopher Karl Popper.[4] To be scientific, the paradigm must be subject to falsification. It must be testable in such a way that its tenets can be examined and reexamined by skeptical scrutiny. Such a view is quite the opposite of faith-based paradigms, which are all basically dogmatic. Scientific paradigms are expected to yield an improved, hopefully predictive, and pragmatic understanding of how the universe really works. Also, scientific paradigms are not culturally specific but cut across cultural distinctions. There is no such thing as Muslim chemistry or Christian biology or Hindu physics. Quantum mechanics, plate tectonics, or any other scientifically based paradigm applies equally to all races, religions, and ethnicities. That is to say, it has no bearing whatsoever on anything cultural. It merely is.

It is true that science, and more particularly scientists, give up cherished paradigms with great reluctance, and such reluctance eventually can lead to what are called scientific revolutions, such as when relativity replaced Newtonian mechanics as a prevalent paradigm. Einstein was loath to admit that his own work on relativity theory indicated an expanding universe, as confirmed by Edwin Hubble's observations and later understood as resulting from the big bang. Einstein, who said, "God does not play dice with the universe,"[5] preferred to think of the universe as in a steady-state, indeed, "balanced." But Einstein nonetheless did eventually and graciously accept the conclusion that the universe is expanding.

It is also true that scientific paradigms historically have clashed with cherished socio-religious or political paradigms. Such clashes are not scientific revolutions in the strictest sense, but go beyond, affecting the social fiber. To cite perhaps the two best-known examples, Galileo's confirmation of Aristarchus's and later Copernicus's concept of a heliocentric rather than geocentric solar system certainly did not win him many friends among the Roman Catholic Church hierarchy for the obvious reason that such confirmation appeared to directly challenge and, in fact, refute the teachings of the Church, teachings that were presented as infallible. Darwin's mechanistic theory of natural selection remains anathema to fundamentalist Christians. When the populist William Jennings Bryan helped prosecute John Scopes for teaching the doctrine of evolution, Bryan's objections to Darwin's theory were not scientific. He could not even accurately describe the theory. Bryan objected to what he believed were the implications of Darwinism, thinking that such notions as humans evolving from apes could, if widely held, undermine faith, at least in the literal interpretation of biblical writings, and thus undermine the moral teachings of the Bible as they act to hold together society.[6]

Scientific paradigms, at least in the modern sense, confine themselves entirely to the material world with no supernatural or teleological components, an essential distinction from other paradigms, many of which have been historically far more dominant in human thinking. It is readily apparent that over the course of human history metaphysically based paradigms, as belief systems for understanding the workings of the world and universe, have been increasingly discarded in favor of materialistically based scientific paradigms. The prevalent scientific paradigms by which educated people understand the world and universe today include the heliocentric Copernican view of the Solar System (including the nebular hypothesis of Leplace and Kant for the formation of star-planet systems), the Darwinian view of organic evolution, atomic theory and quantum mechanics, Einstein's theories of relativity, the concept of spacetime, the big bang, and subsequent expansion of the Universe, and plate tectonics and continental drift. These paradigms, none of which includes a balance of nature, greatly overlap with one another and together provide us with the scientific worldview.

As I shall describe, the balance of nature, as paradigm, originated as a metaphysical, teleological concept, though some ancient Greek philosophers did attempt to view it by way of what would be called today a scientific approach. Within the various modern scientific paradigms, the balance of nature concept does not really stand alone but is included, tested if you will, largely within the paradigm of Darwinian evolution.

The balance of nature paradigm is of little value within evolution and ecology. It has never been clearly defined and is basically misleading. But the balance of nature is esthetically satisfying, a fact that is largely responsible for its continued vigor through the ages.

Scientific paradigms, as described above, do not automatically contain an esthetic component. Nor are they necessarily intuitive. Humans see the world through limited sense organs (we do not see in the ultraviolet color range or hear very high frequencies, for example), and senses can be deceiving. Most scientific paradigms are actually counterintuitive, sometimes remarkably so. Organisms, at least over a few human generations, don't appear to evolve. The Sun does appear to move across Earth's sky in the course of a day, an artifact, of course, of the daily rotation of the Earth about its axis. Probably fewer than one percent of the persons on Earth understand Einsteinian relativity, either special or general, well enough to attempt even a crude explanation of it. A detailed knowledge of quantum mechanics is well beyond the understanding of most people. But comprehending a scientifically based paradigm still does not assure its societal acceptance. Darwinian evolution is fundamentally quite easy to comprehend, at least at a basic level, and it is still widely disregarded by many otherwise educated people. Why is this?

Scientific paradigms are not always easily learned and do not automatically yield a sense of personal satisfaction with one's place in the big scheme of things. In fact, the opposite is usually the case. Scientific paradigms, when seen in the larger context of societal history, tend to reduce imagined or wished-for human stature in the universe. This apparent diminishment may be why so many otherwise well-educated people fail to embrace science as a way of knowing the world and the cosmos. Indeed, some brag about their scientific ignorance, as though doing so somehow excuses it. It is hard enough, for example, to comprehend the universal inflation that presumably occurred at the initiation of the big bang (when all that is the known universe had been compressed into a singularity less than the size of an atom), let alone develop a warm and fuzzy feeling about it. Knowing that the position of an electron in an orbital shell of a magnesium atom can never be known for certain may or may not seem disconcerting. Being told that there is really no balance of or purpose to nature or that your distant ancestor was an ape perhaps hits closer to home and thus is more likely disconcerting. But disconcerting or not, the above statements appear to be true.

The relationship of the human species with the rest of nature has occupied thinkers ever since there were thinkers. Surely one of the first queries that would deserve to be termed "philosophical" must have been the consideration, by prehuman hominids, of the other creatures with which they roamed African savanna ecosystems. What were those varied, diverse-looking animated beings so very different from themselves? Some, it was quickly learned, were clearly dangerous, to be avoided at all costs, while others were not. Some were numerous, clustered together in large groups that daily could be found roaming across the vast grassland, while others were solitary and rarely seen. What did our temporally distant ancestors think about antlers, fur, teeth, claws, and the various beasts that possessed them? Did they contemplate how and why some animals seem inconspicuous against their backgrounds while others are much more obvious? Did they ever wonder, even as they feared them, why the animals we now call leopards were so much less numerous than the animals we now call gazelles? How did they regard trees and grasses? Were these plants fundamentally different from rocks and soil, alive, as we know them to be today, or were they only vaguely distinguishable among the many characteristics of the land that were other than animals? Most importantly, when did our hominid ancestors first see themselves as uniquely distinct from the other animals? We'll probably never know.

Who were the first members of our lineage to know themselves as individuals? Was it an *Australopithecus afarensis* who first gazed into a quiet pool of still water and realized just what that reflection was? Precise self-awareness was one of the first essential intellectual steps to be climbed as consciousness gradually emerged and natural selection expanded the prehuman brain to become the human brain. Nonhuman animals such as domestic cats, for example, are quite egotistical, with a strong instinct for self-preservation and self-maintenance, hunting, eating, grooming, and resting. Such creatures have robust memories, strong personality traits, a high capacity to learn certain kinds of information, and numerous subtle nuances of behavior that identify them not just as cats but as individuals. Anyone who shares his or her life with animals knows how individualistic animals really are. But does a cat know who it is? When a cat gazes into a mirror, does it know that it sees its own image? Humans do. It may have been one of the Australopithecine species, *afarensis* perhaps, or maybe *africanus*, it may have been *Homo habilis* or even *Homo erectus*, but surely one of these ancestral hominids that preceded the appearance of *Homo sapiens* acquired the profoundly important trait of self-recognition. Largely instinctive egotistical behavior is basic in the world of animals, and hominids were certainly no exception. But in the case of early hominids, somewhere in time some neuronal connections synapsed in such a way as to provide an actual sense of self. How extraordinary.

To borrow a phrase from Carl Sagan, that was "a very big thought."[7] The awareness of self may be considered the first paradigm, a view of the world limited to the notion that "I am me."

With the intellectual capacity for self-recognition also came the ability for recognition of others, again a trait not in any way confined to hominids. Animals recognize others of their species as well as other species—goats recognize horses, for example, just as cats and dogs recognize specific humans and know them to be different from themselves. Cats hiss at one another but they run and hide from large dogs. Dogs sniff each other as they become acquainted and they predictably chase squirrels at any opportunity. Human beings, as well as other primates, are genetically and ecologically social animals, with a long evolutionary history of adapting to living within a group. Indeed, such group dependency is so deeply ingrained in the behavior of humanity that one of the most brutal punishments that can be inflicted upon a human being is solitary confinement, the deprivation of virtually all human contact. With the intellectual capacity to know each other as individuals, social groups exhibit a pattern of behavior more complex, with greater interdependency, than that of solitary species.

Those social animal species with an intellectually well-developed brain learn each others' identities with great precision and act in ways that bring personal rewards in exchange for various forms of gratuitous behavior directed toward others in the group. Mutual grooming, a common practice in human, chimpanzee, and baboon societies, is a form of "reciprocal altruism," where individuals remember exactly who picked ticks from them and return the courtesy in some form. Both parties benefit, though the interaction is not always totally symmetrical. Sometimes one benefits more than others if it succeeds in getting groomed more than it has to groom others. Human societies have obviously taken reciprocal altruism well beyond tick removal. Indeed, it is essentially the basis for economics.

For humans, the evolution of what Darwin called "the social instincts" coupled with the emergence of a brain sufficiently complex to store, rearrange, and quickly process vast amounts of experiential information, may have represented the second paradigm in the history of our species, the belief that not only "I am me," but also "I am part of my group."[8] Such recognition that one's self is bound socially with one's colleagues probably accounts for the xenophobic tendencies that are typically exhibited by one social group toward another, or by members of a group toward a stranger or strangers. It also stimulates strong cooperative interactions among members of the group. But strong self and group identity goes further, mentally setting the identity group fundamentally apart from all of the rest of the natural world. This characteristic represented an intellectual foundation for what would become a dominant paradigm of humanity, the dualism between humans and all else in nature. That step in awareness of self, group, and species is really when humans first conceived of what may accurately be called "nature." Nature was not us.

Nature could punish or reward, provide shelter and sustenance, or change unpredictably. Nature was complicated, mysterious, scary, and fickle. At the same time it provided sunny warm days, cool star-filled nights, beauty, and, to be sure, everything that was food and shelter. Perhaps of singular importance, nature was not easy to understand. The natural world abounded in mysteries, and was full of surprises, not all of them pleasant. As human intelligence was emerging, nature, including solar eclipses, monsoons, drought, predators, poisons, pathogens, and insect hordes, must have posed intellectual and emotional challenges unlike anything a person of the twenty-first century can really imagine. How can one possibly make sense of such events? The world must have seemed very capricious back in the earliest days of humanity. It is easy to see why gods were called upon to explain so much that was then otherwise unexplainable.

As the evolving human intellect was coping with and attempting to comprehend nature, it was also developing another profoundly important attribute. It was becoming fully aware of death. Humans quickly learn that other humans die and that such events immediately deprive them of any further contact with persons whose company was once of great satisfaction to them or upon whom they depended. Grief, as with love, is a paramount human emotion. Even more importantly, individuals, like it or not, come to understand that they are each mortal. Death eventually claims all persons. This thought is even bigger than "I am me." In the 1992 film *Unforgiven*, the lead character, a gunman named William Munny (played by Clint Eastwood), said it very well: "It's a hell of a thing, killin' a man. You take away all he's got an' all he's ever gonna have."[9] That realization goes a very long way toward accounting for why societies developed religions, virtually all of which include a belief that some essence of the individual must persist after death. The concept of the soul or afterlife likely appeared in human culture as human consciousness emerged during the evolutionary hypertrophy of the frontal lobes of the brain. The soul, the spiritual self, links the human psyche with the belief that there is purpose to one's existence. It is quite reasonable to want to believe that one's life has some profound purpose to it. And it can easily be said, "If I have purpose, does not all else have purpose as well?" So a belief in a personal purpose for existence, a reason for being, leads easily to the belief that nature must have some purpose as well. Teleology is the philosophical belief that life has purpose and was designed with purpose. It is really a form of human egoism applied to all of nature and the rest of the universe as well. It is a form of pandering to an ego that demands worth and purpose out of the temporary event that is life.

The realization of the inevitability of death and the need to believe that something is attainable beyond it, coupled with hoped-for relief from pain and suffering, plus the absolute mystery and awe in which nature must have been held by early tribal groups of humans, combined to impart to human intellect a strong tendency toward spiritual belief, a spirituality so important as to have become the dominant driver of world paradigms. Life, as has been noted, is dangerously unpredictable. How can such unpredictability be explained, rationalized, accounted for in such a way that the human psyche remains intact? Belief in a spiritual world, a "better" world, an ordered and superior world beyond the present one, goes a long way in mitigating the sad realities characterizing life and death. It is not surprising therefore that spirituality is a prevalent component of the human mindset. Indeed, it is hard to imagine any emerging intelligence anywhere in the universe that would not at least pass through a "spiritual phase" during the maturation of its intellect. The concept of a deity, a god or gods, provides a way of uniting the physical with the spiritual, supplying both balance and purpose. Any deity would be seen as ordered, not disordered, and the form taken by the order would, having been created or imposed by God, surely have purpose. If nature is ordered, nature, and presumably all in it, has purpose. The order believed imparted to nature by the Creator is what would become known, largely by assumption, as the balance of nature.

It is not surprising that paradigms that developed within ancient cultures, including the paradigm of nature's ultimate balance, historically have spiritual, metaphysically based roots. By the time early agriculture was developed about 10,000 years ago, the human brain had been its present size and neural sophistication for at least 50,000 or more years. As societies formed, as city-states emerged, and as language developed, cultures assimilated an overarching spiritual view of the world to help them cope, organize, and govern their lives. Nature was obviously very much a part of spiritually based belief systems. In many cultures, understanding the flood cycle, for example, was of crucial importance in knowing when to plant crops. Practices such as human sacrifice often were employed in the hopes of satisfying a presumed spiritual overseer of nature, an extreme example of reciprocal altruism where human life is exchanged for omnipotent intervention in hopes of assuring meteorological satisfaction.

Knowledge about nature was originally part of oral cultures. Presumably the written word (or symbol) appeared long after language developed. The oldest writing is barely more than four millennia old. Language existed long before then. Even today, cultures exist that have little or nothing in the way of writing but that have a rich tradition of oral history. Such cultures, some of which can be found among Amazonian indigenous tribes, typically demonstrate a sophisticated understanding of local plants and animals. Ethnobotany, for example, is the study of how tribal peoples use extracts from numerous plants to manufacture drugs, poisons, fiber, etcetera. Typically much of the knowledge rests with the shaman, an individual who uses a combination of pragmatic techniques and

magical practices designed to involve the spirit world as components of the rituals by which powerful drugs are extracted from
rainforest plants. A shaman sees little difference between purging
intestinal worms and purging a demonic spirit from an individual.
Indeed, both objectives could be part of the same ritual process.
Nature is understood to be spiritual as well as literal, and none of
the pragmatic information about nature in any way overrides the
spiritual forces that lurk within the natural world. Rituals of ap-
peasement normally accompany treatments. A shaman would likely
consider the ritual component of his or her job at least as important
as pharmaceutical skills.

Oral traditions do not lead to science in any historical sense. It
is too easy to forget or change words when only memory allows
their recall. And it is very hard to do mathematics without writing
it down somewhere. Given how importantly math skills factor into
most of science, it becomes obvious that science had to await the
written word. The skill with which shamans do their work demonstrates that science is not the same as technology. It is possible to
invent technology by trial and error (as in the precise, meticulous
extraction of curare from certain vines), but doing so does not automatically lead to broad scientific understanding. Science did not
appear in any meaningful way until the invention of writing.

It was in ancient Egypt and Babylonia where the first human
thoughts were transcribed on tablets, papyrus, or some other vehicle such that they were preserved beyond mere memory. Picto-
graphs, hieroglyphs, and symbolic figures came to represent combinations of sound that conveyed information. Alphabetic writing
was invented presumably by the ancient Greeks (from the Phoeni-
cian alphabet) approximately 800 BC; some scholars believe this
was the single most pivotal event in what would be the emergence
of philosophy and science. Now it was possible to compare thoughts
among people in a precise way: they were written down!

The ancient Greeks traveled widely and acquired knowledge of
the mathematical work of the Egyptians and Babylonians. At that
time astrology and astronomy were still one and the same. Studies
of the patterns and movements of star groups compared with the
"wanderers," the planets, led to the invention of the Zodiac, the

twelve constellations along the ecliptic plane, the plane of the Solar System through which the Sun passes in the course of a year. Astrologers coupled mathematical studies of the sky with predictions of dubious merit about how stellar and planetary configurations would ultimately influence lives and events. What is most remarkable about this historical marriage between reason and mysticism is that it still persists, indeed thrives. But the example serves to show that even as sophisticated knowledge of the movements of stars and planets was emerging, that information was used at least in part merely to buttress baseless beliefs, paradigms of spiritual influence. The reason is obvious. Those who understood the movements of the stars were the priests of the various cultures, individuals of unique privilege and influence. Their high status in the society was based on their presumed abilities to interface with the powerful spirits whose contentment or wrath would, it was believed, determine if things would go well or badly. There was no point in attempting to segregate objective analysis from mystical interpretation. It was a politically risky act, potentially even suicidal.

There are simply no records of how cultures perceived nature prior to the invention of writing. Some cultures must have suffered plagues, for example, which would seem obvious examples of how nature can become stunningly imbalanced. But most early cultures believed a plague to be something visited upon them for failing to adequately serve or honor their gods, and thus they saw the plague as a form of direct intervention by the gods specifically to punish them. With some combination of prayer, sacrifice, and ritual, the gods could hopefully be pacified and the balance of nature would be restored.

With the invention of alphabetic writing, the ancient Greeks emerged as the first philosophers and scientists. Ancient Greek thinkers were diverse in their beliefs, but one thing that they all attempted to do was to bring order to understanding the natural world. The word "cosmos" comes from the Greek *kosmos* for "order," just as "chaos" is from the Greek word for disorder. Ancient Greek thinkers converged in their efforts, lasting over hundreds of years, to wrest as much information from the chaotic world and place it firmly within the world of the cosmos. Earth, air, fire, and, water, the four basic components of the Greek cosmos, were viewed by various Greek philosophers as being in interactive balance, perhaps the initial expression of what could be thought of as the grand balance of nature.

Greek philosophy developed by approximately 600 BC, and what emerged over the course of the next few hundred years was what amounted to a struggle for dominance between teleology and materialism. Materialism is the belief that the universe operates by a series of natural laws that are knowable and thus allow a rational understanding of phenomena, one not based on assuming design or purpose. Teleology, which was more philosophically appealing, eventually "won," at least to the degree that civilizations following ancient Greece perceived nature and the universe in highly teleological terms, citing above all others the works of the greatest of all teleologists, Aristotle. The dominance of what was basically an Aristotelian cosmology and the Aristotelian view of nature would last well into the seventeenth century. This was regrettable for science in general and for evolutionary biology and ecology in particular. And Aristotelian influence persists. It is a principal reason why many today still believe in at least some measure of teleology, and even more subscribe uncritically to the belief in some kind of balance of nature.

Materialism arose and flourished in ancient Ionia, in what is now the country of Turkey, located on Asia Minor across the Aegean Sea from Greece. It was here that such thinkers as Thales, Anaximander, and Democritus, to name but three, became focused on a profoundly important approach to knowing the world: the belief that the world is, in fact, knowable by investigation. The roots of what we now call science were planted in such places as the island of Samos and the city of Miletus. The Ionians were materialists in the sense that they investigated the phenomena of the world without relying on the whimsical behaviors of the twelve gods of Mount Olympus, the gods of Homer and Hesiod, as causal factors in the daily workings of the planet. Anaximander, in a description that bore an evocative similarity to the big bang, mused that the universe sprang from a small seed contained in the *apeiron*, a vast and boundless chaotic space. Democritus formulated the atomist movement, the belief that life and all else was composed of atoms (units that could not be subdivided into smaller units) behaving machinelike, according to natural rules that had nothing to do with divine influence. This highly mechanistic perspective included the notion not only that the atoms behaved following strict, invariable rules, moving about in space, but that the world of matter composed of atoms had no intrinsic purpose. It just existed. Democritus did not require the existence of a soul. An atomist view of nature more closely resembled what would become the Darwinian paradigm than it did the Aristotelian paradigm that would eventually predominate Hellenic and post-Hellenic thinking.

The Ionian view of natural laws still allowed for the cosmos to be in balance, to have an internal harmony. The forces that governed the physical, and presumably the biological world, attained a natural balance in Ionian philosophy. Some Ionian thinkers believed that such harmony ultimately may be due to the action of the gods, but once put into place, the gods would no longer intervene and tinker with their creations. What is important about the Ionian approach, its real legacy, was its belief that there are laws to govern nature, natural laws, and these materialistically based laws of nature are knowable by actual investigation.

This rationalistic approach to acquisition of information spread rapidly throughout the Greek empire. Anaxagoras, who lived around 450 BC, taught that the Sun and other stars were distant orbs, giant rocks hot with fire, and that the light of the Moon was actually reflected sunlight. Eratosthenes, born around 276 BC, used trigonometric methods to produce an elegant measurement of the circumference of the Earth accurate to within a few percent, a remarkable achievement for that time. Hipparchus, born in Ionia circa 190 BC, worked out a mathematical model for the motion of the Sun (still presumed to orbit the Earth) that Ptolemy later used unaltered in his monumental (and incorrect) *Almagest*. And finally there was Aristarchus, born on Samos around 310 BC. He used Pythagorean trigonometry to calculate the relative distances of the

Moon and Sun from the Earth. It was Aristarchus who suggested that the Earth is actually one of the wanderers, a planet, and that it orbits the Sun, not the other way around. The paradigm of a heliocentric rather than the geocentric system is often credited to Copernicus, but Aristarchus, an Ionian, first conceived it.

Physical science dominated Greek philosophy, but biological science was not entirely neglected. Biological research dealt, understandably enough, mostly with medicine and physiology. Again, as with the physical sciences, the overall conclusions were that physiology was fundamentally knowable by observation and investigation. Gods were not seen to be the causal factors for plagues or disease, and medical treatments, such as they were, did not focus on the use of magic. The primary paradigm for health was that the human body functioned through a balance of bodily humors (blood, phlegm, yellow bile, black bile) and that various maladies could be cured by restoration of the proper balance. As with the rest of Greek thinking, attaining balance was essential. Of the various practitioners, Hippocrates, born around 460 BC, ranks as most influential. Under Hippocratic medicine, actual physical examinations were performed and attempts were made to develop treatments for an array of common ailments. Much later, Galen would continue and refine the Hippocratic approach, though Galen would, as will be seen, approach medicine in a remarkably teleological manner.[10]

Nature, in ancient Greece, was generally assumed to be unchangeable, a kind of steady state, in keeping with the notion that the cosmos must be balanced. Such a belief was likely based on nothing more than intuition. No studies on nature were performed that could be called equivalent to the detailed measurements of Eratosthenes, for example. Aside from Aristotle's efforts, observations of nature were casual, though sometimes raising significant questions. Herodotus, who was a historian of the fifth century, commented on an issue that would later intrigue Darwin, the fact that different species of animals have widely different fecundities. Noting that predators don't consume all of their prey, Herodotus assumed that Providence had provided for greater reproductive output among prey animals (such as hares) specifically to protect them from being entirely consumed by predators, which, Herodotus noted (presumably with satisfaction), typically have lower reproductive rates. Herodotus's view of the balance between an animal's position on the food chain and its reproductive capability, as assigned it by God, sowed the seeds of what would be called "natural theology," a view of nature that would predominate in European culture right up to the time of Darwin. The concept of "intelligent design" that is so popular now is merely a repackaged version of natural theology (chapter 4). Herodotus also noted the existence of mutualisms in nature, observing, for example, that a Nile crocodile will permit a certain plover species to forage in its open mouth, the bird extracting leeches from the reptile, both animals benefiting in the process. Such anecdotal observations provided good fodder for the notion that nature was well planned and nicely balanced.[11]

Socratic philosophy differed significantly from the Ionian approach. Rather than focus on learning the workings of the natural cosmological world, Socrates (470–399 BC) concentrated on the underlying philosophical bases for ethics and political decisions. This approach led to the use of logic and argument to ascertain what, for want of a better word, could be called "truth." The famous "Socratic method," employed by countless college professors in the ages that followed, consists of persistent, probing questions, an attempt to establish what is called a Socratic dialogue. Ionian experimentation and analysis was a very different methodology from Socratic analytical thought. Without question, one of the best of the Socratic thinkers was Socrates' student Plato. Plato (429–347 BC) did not confine his philosophic work to ethics and politics, but considered also, in his dialogue the *Timaeus*, the world of nature and the cosmos.

Plato rejected any Ionian notion that the cosmos was a mere atomistic collection of things that responded only to their own intrinsic natures, somehow achieving order. As the historian David C. Lindberg has written, Plato was "convinced that the order and rationality of the cosmos could only be explained as the imposition of an outside mind." Thus Plato believed that the universe, and all in it, had both plan and purpose, an order imposed upon it by what he called the demiurge, or divine craftsman.[12]

In Plato's philosophy the soul was basic, the body secondary. Bodies were mere imperfect essences of ideal, unchangeable forms. I sometimes illustrate this point by asking my class who among the men would care to stand up as an example of the "ideal male human," and who, among the women, would rise and characterize herself as the "ideal woman"? I usually get no takers. Plato would tell the students that they are each an imperfect copy of the *eidos*, the ideal human form, that exists somewhere beyond earthly reach. The famous Platonic allegory of the cave makes the point. Plato described how people imprisoned in a cave and able to view only the shadows of persons or objects against the wall of the cave would eventually come to see those shadows as "reality," when they are, of course, mere shadows. In a philosophically like manner, our everyday experience is to see but imperfect representations of what exists elsewhere in perfect form. These forms were carefully crafted by the demiurge and they are changeless and ideal. Plato's philosophy of essentialism, as has been pointed out by the evolutionist Ernst Mayr, was unabashedly creationist, leading Mayr to describe Plato as "the great antihero of evolutionism."[13] Variation was accounted for in Platonic thinking by trivializing it. All species were created as perfect essences represented on Earth by imperfect forms. Such imperfections are of no great significance. If species were created ideal and changeless, then combinations of species, what we call nature, must be likewise.

Platonic nature could not be other than balanced. Any appearance to the contrary would be of no great significance, as Plato taught that true knowledge was not attained by mere observation. In other words, if it seems to be imbalanced or changing, ignore it. Frank N. Egerton has pointed out that in the *Timaeus*, Plato suggests that the various species of the Earth are all akin to organs in a kind of super-being or supra-organism (see chapter 6), a concept of a balanced nature that remarkably remained vibrant in ecology throughout much of the twentieth century![14]

Without question, Aristotle (384–322 BC), Plato's prize student, was the most significant of the Greek philosophers in the field of biology. His intense, detailed observations of nature developed into a paradigm that lasted for two millennia, and, to a minor extent, his influence continues to the present day. Aristotle, though not an experimental biologist, made thorough studies of marine and other animals, including meticulous dissections. His work in zoology was published in several large volumes that survive to the present day, including *On the Parts of Animals* and *History of Animals*. In putting together the vast amount of biological information, Aristotle developed an elegant and satisfying view of nature. It is completely wrong.

Aristotle viewed balance as perhaps the single most important component of the cosmos. His belief in an eternal, full, balanced universe was directly applied to nature as well as astronomy. Beyond that, he accounted for change in the cosmos (whether on Earth or beyond Earth) as the product of four possible causes: formal cause, material cause, efficient cause, and final cause. For an understanding of Aristotle's view of nature, final cause is the most important of the four. Aristotle meant it to be teleological: that the world is shaped and structured by purpose, things happening in an orderly manner, their natures guided by a divine plan for the universe. Aristotle's view of nature was obviously in direct opposition to the atomists, who viewed chance as having a major role in shaping things. Aristotle was unsympathetic to any notion that the universe was so influenced.

Aristotle strongly believed in plenitude, the notion that the universe and, by implication, all of nature, is as full (and balanced) as it can be. What this means is that each habitat typically contains the number of species of plants and animals that it is designed to contain. It is clear from Aristotle's definition of causation that plenitude also implies that nature is intrinsically balanced, all creatures from pansies to peacocks present for a specific purpose, to play a fundamental role. Centuries later, this component of the Aristotelian paradigm of nature would pose a particular challenge to Darwin in explaining evolution.

What Aristotle is best known for is his complex, hierarchical arrangement of animal and plant species based upon their presumed degree of perfection and the level of development of their souls.

Aristotle said that only humans possessed a "rational soul," so humans occupied a high position on the hierarchy that is nature. This hierarchy was an ordering leading from least to most perfect, with what Aristotle called the Prime Mover at the pinnacle. Males occupied higher positions on this *scala naturae*, or "great chain of being," than did females. Monkeys were higher than dogs but dogs were higher than lizards. Plants, with merely a "nutritive soul," were near the bottom.[15]

No creature can ever change position on the great chain. To do so would violate its essence and imply that its purpose had been altered. Such alterations would be too profound to find acceptance in Aristotle's scheme for the perfection, balance, and design by which nature is organized. Some interpreters of Aristotelian biology have suggested that the great chain hints of evolution but, as Ernst Mayr has clearly shown, just the opposite is the case.[16] It is totally creationist and teleological. From his detailed studies of embryology, developmental biology, anatomy, and natural history, Aristotle was unabashedly committed to the view that all of nature had a guiding purpose to accompany its function, a cosmos where interrelationships among species are part of the grand balance that is life, indeed, that is nature.

It would be an understatement to say that Aristotle had great influence on future thinkers. He firmly established Platonic essentialism and teleology as the paradigm for nature. Centuries later Galen, an extreme teleologist, articulated how the human body is perfectly designed to meet all its needs, a clear extension of Aristotelian philosophy into medical science. Aristotle's paradigm of a directed, purposeful, balanced but static nature endured for 2,000 years until evolutionary biology began to undermine it. Even today, contemporary historians continue to view Aristotle's paradigm with more sympathy than it deserves. David C. Lindberg, in an account of Aristotle's science, remarks that "it is nonetheless worth noting that the emphasis on functional explanation to which Aristotle's teleology leads would prove to be of profound significance for all of the sciences and remains *to this day* a dominant mode of explanation within the biological sciences."[17] Ernst Mayr, while noting that Aristotle's philosophy was a severe impediment to the development of evolutionary thinking, does credit him with one very significant accomplishment. As Mayr put it, "Evolution, as we now realize, can be inferred only by indirect evidence, supplied by natural history, and it was Aristotle who founded natural history."[18]

Ecology eventually followed.

∾ 4

Ecology B.C.
("Before Charles")

When did ecology, the science, really begin? Ecology is rooted in the Greek word *oikos*, referring to home, and the word was not invented until the second part of the nineteenth century, when the German zoologist Ernst Haeckel[1] used it in 1866, seven years after Charles Darwin published his most famous and influential work, *On the Origin of Species*. Haeckel meant the word to mean roughly the "household of nature," or the "economy of nature," a term Darwin had used in the *Origin* in his description of natural selection. Darwin wrote,

> I am convinced that the whole economy of nature, with every fact on distribution, rarity, abundance, extinction, and variation, will be dimly seen or quite misunderstood. We behold the face of nature bright with gladness, we often see superabundance of food; we do not see, or we forget, that the birds which are idly singing round us mostly live on insects or seeds, and are thus constantly destroying life; or we forget how largely these songsters or their eggs, or their nestlings, are destroyed by birds of prey; we do not always bear in mind, that though food may be now superabundant, it is not so at all seasons of each recurring year.[2]

What is of importance here is that both Darwin and Haeckel recognized how essential it is to study organisms in the context of their abiotic (nonliving) and biotic (living) environments. Implicit in this view was that nature was "balanced" and thus the economy of nature was the formal study of such balance.

The science of ecology came into being only a little over 130 years ago. But its actual history goes back considerably further. I call this chapter "Ecology B.C. ('Before Charles')" because the study of nature was a very different enterprise prior to November 24, 1859, the day upon which *On the Origin of Species* was published. Once Charles weighed in, things were destined to change.

Science was largely quiescent throughout the time period extending from the fall of Rome to the emergence of the Renaissance, though there were a few scientific thinkers ("rationalists") as early as the twelfth century.[3] Had Aristotle, who, as I discussed in the last chapter, was the founder of natural history, been able to talk with any learned person of, say, the thirteenth century, over a millennium after he died, the legendary Greek philosopher would have learned little new about science. Medieval times were good for knights and monasteries but bad for science, a reality reflected in the popular name for the period, the Dark Ages.

Beginning roughly in the fifteenth century, scientific study, inquisitiveness about the workings of the world, was slowly reborn in Western culture, a time of renaissance, the emergence of the modern worldview. For at least a few folks, curiosity got the better of them and this thing we call "science" got rolling. What really stimulated science was the developing relationship between science and technology. Humans began learning how to invent and use labor-saving machines to extract services from nature. Human anatomy, physiology, and the many infirmities of humanity were slowly investigated. Science, particularly the physical sciences, emerged. Science happened as a result of curiosity and because scientific knowledge ultimately paid dividends.

This was the vibrant, exhilarating time of Leonardo da Vinci and Nicholas Copernicus, when the intellectual stage was set for what would become, over the course of the next centuries, a continuing series of scientific revolutions.[4] These intellectual adventures would profoundly change our view of the cosmos and the place of humans in it, as well as lead to the development of ever more complex technologies that make our modern lives immensely more pleasurable, healthy, and safe. Imagine how the modern world would appear to someone from the Renaissance. As the science fiction author Arthur C. Clarke often noted, it would all be magic!

Curiosity about the world was also related to imperialistic economic ventures, the historic voyages of exploration that revealed a world much vaster and more biologically diverse than previously imagined. Explorers such as Marco Polo, Ferdinand Magellan, Christopher Columbus, and Vasco da Gama, as well as numerous others, brought back fascinating examples of animals and plants that were previously unknown to Europeans. These included examples of the great apes—gorilla, chimpanzee, and orangutan—which really stimulated interest. The first to be discovered was the orangutan, from tropical Asia (especially Borneo) in 1778. This was followed by the chimpanzee in 1788 and the gorilla in 1847. Apes created a sensation in more ways than one. Their brutish appearance was obvious, but, on the other hand, their undeniable resemblance to humans in general body form was disconcerting to many.[5] Darwin was quite spellbound by a young female orangutan named Jenny who resided in the London Zoo. His observations of Jenny most certainly influenced his thinking on human origins. As similar as great apes are to humans, it would nonetheless be over three centuries until George Schaller, Jane Goodall, Dian Fossey, and others would document the sentience of the great apes and at least some measure of humanity would come to regard them, as Darwin argued, to actually be our nearest evolutionary relatives.

The imperialistic voyages of the eighteenth and nineteenth centuries made this time period the "great age of natural history." Biology was born, more or less. It became customary, especially on British voyages of exploration, for a naturalist to join the ship's company to collect, describe, and organize the new species of plants and animals that were inevitably found. On his first voyage, Captain James Cook, who subsequently explored much of the tropical Pacific Ocean (and discovered the Hawaiian Islands), brought along a botanist, Joseph Banks. Banks later made immense contributions to understanding the diversity of the tropical plant world. In London, the British public was fascinated by the diverse animals from distant continents that were displayed in zoos and museums, the trophies of the East India Company and other voyages of exploration and commerce.[6]

Alexander von Humboldt, who arrived in South America in 1799 and traveled extensively throughout the continent, not only wrote in detail about the lush lowland rainforest but also detailed the various ecological zones encountered when moving up the slopes of the Andes mountains.

Of even more significance, Captain Robert FitzRoy, then the youngest commander in the British Navy, selected the youthful Charles Darwin, freshly graduated from Cambridge University, to accompany the ship HMS *Beagle* on a voyage around the world, with particular emphasis on South America and the Galápagos Islands. FitzRoy, upon first meeting Darwin, was initially troubled by the rounded, gentle shape of Darwin's nose. FitzRoy was a believer in phrenology, the notion that the shape of one's head reflects one's character. He feared Darwin's nose reflected a weak character. In his autobiography, Darwin wrote of FitzRoy, "he doubted whether anyone with my nose could possess sufficient energy and determination for the voyage. But I think he was afterwards well-satisfied that my nose had spoken falsely."[7]

This trip, lasting from December 7, 1831, to October 2, 1836, was to profoundly change biology. It was on the voyage of the *Beagle* that Darwin matured as a naturalist and scientist, making exhaustive observations and notes. He collected an array of organisms that, once he returned to England, formed the basis for his brilliant insights into evolution, the theory of natural selection, and, ultimately, what would become the study of ecology. But we'll delay talking more about Darwin until the next chapter.

As more and more kinds of plants and animals were being discovered, thinkers sought ways to effectively organize this vast newfound wealth of information. Voyages to distant lands and travels around Europe kept adding to the lengthening list of what was then called "the Creation." Why the Creation? Because merely finding more and more forms of life did utterly nothing to alter the paradigm of creationism or the assumed balance of nature. That scientific revolution had yet to occur.

The invention of the microscope by Anton van Leeuwenhoek in the seventeenth century showed that myriads of plants and animals, previously unknown, were present in the world, but far too small to be seen with the naked eye.[8] This, of course, should have made some wonder about why a Creator would make so many things we could not see, but, then again, apparently the subject was not philosophically troubling at the time. It certainly was not troubling to Leeuwenhoek. He was a Dutch Reformed Calvinist, and whatever he discovered swimming around on his microscope was simply further proof of the omnipotence and omniscience of God's work. This included bacteria, which he took from human mouths, discovered in 1676, and sperm cells, which he observed and described in 1677.

And it wasn't merely living, extant nature being discovered. The fossil record was being revealed in increasing detail. What, exactly, were these impressions in the rocks? Beginning with careful studies by Robert Hooke in 1665, fossils were finally regarded as remains of previously existing life forms. That was somewhat troubling to the firmest believers among creationists. Geologists were learning that ancient environments bore little similarity to those that currently exist. This raised questions. Seashells in sediments high on mountainsides? Impressions of marine fish from rocks found in inland arid regions? Trilobites? What were they? Beyond that, geologists, stimulated by the efforts of James Hutton, were getting hints that the age of the Earth was profoundly greater than the commonly accepted 6,000-year estimate associated with Christian theology. As Hutton described the Earth, "No vestige of a beginning, no prospect of an end."[9]

By the time the first of the dinosaurs, a carnivorous species subsequently named *Megalosaurus* (an animal somewhat like *Allosaurus* but smaller), was discovered in 1824, it had become clear that life's panorama in time was older and stranger than had been previously believed. Credit the brilliant French anatomist Georges Cuvier for recognizing that numerous animal communities, now extinct, had existed in the distant past.[10] James Hutton in Scotland as well as other geologists were gathering data clearly showing that Earth was far older than previously believed. Because of

Cuvier's efforts, the concept of extinction gradually came to be accepted. After all, for those clinging to the strict doctrine of creationism, the Flood could have done them all in (though you do wonder how so many fish and clams drowned).

Prior to Cuvier, many educated people did not believe in extinction, among them Thomas Jefferson, who logically assumed that a perfect Creator would not make life forms so imperfect as to cause their ultimate demise.[11] Jefferson clung to the belief in the *scala naturae,* and saw no reason why unique fossil animals could not still be thriving in some distant, as yet unexplored part of the world. He likely hoped that Lewis and Clark would find a few on their famous expedition. But when the intellectual and fossilized dust settled, Cuvier showed beyond doubt that many species not present on Earth now, once were. Extinction was real. The first stirrings of serious evolutionary thought were out and about.

To step back somewhat, throughout the seventeenth and eighteenth centuries efforts were centered on finding a satisfactory schema for the classification of the ever-increasing bounty that was nature. Using the work of Aristotle as a foundation, John Ray authored such works as a *Catalogue of Cambridge Plants* (1660) and his major effort, *Wisdom of God Manifested in the Works of Creation* (1691). Ray's title for the latter work reflects an additional philosophical agenda, that of not only classifying, but interpreting nature as fitting firmly within the structure of Christian theology. Again, implicit is such an effort was the assumption that nature was balanced, that each plant and animal had its role to play in God's created world. Ray's work was really a reincarnation of Aristotelian teleology revised to accommodate the Genesis account of nature. Like Thomas Jefferson, Ray believed that fossil organisms with no obvious resemblance to extant species merely represented undiscovered species that still lived in some remote part of the world.[12] How could God's creation be otherwise?

In 1735, the Swedish naturalist Carl von Linne (whose name is customarily Latinized to Linnaeus) authored *Systema Naturae* (*System of Nature*), an extremely influential work that presented a nested schema for classifying organisms basically still in use today: kingdom, phylum, class, order, family, genus, species.[13] It was also

Linnaeus who developed the custom of calling each species by both genus and species. These names are always in Latin, ensuring that no confusion occurs due to different usage of common names among different countries or regions. For example, "robins" in America are a different species from "robins" in Europe. But both European and American robins have rusty red breasts, and when European settlers first came to North America they saw birds resembling the birds of their native country and called them robins. Because of Linnaeus, robins in America bear the official name *Turdus migratorius* and robins in Europe are named *Erithacus rubecula*, eliminating any possible confusion caused by their common names.

Why was Linnaeus so diligent at his work to classify each and every life form? Like Ray, Linnaeus continued to think of species as each the product of special creation, part of the great balance of nature. Indeed, that belief was inspiring to Linnaeus, and was one of the major reasons for embarking on his life's work of classification.

Linnaeus contributed substantially to what would become ecology. In 1749, a little more than a century before Darwin would publish the *Origin*, Linnaeus published the term "economy of nature." He used the term in a thesis entitled *Specimen Academicum de Oeconomia Naturae*. Darwin, as noted earlier, would later adopt this phrase in the *Origin*. Still borrowing on the basic philosophy of the ancient Greeks, Linnaeus sought to demonstrate that there was a balance of nature in which each and every species of organism has distinct purpose, its own unique role to play. For example, he took pains to credit maggots for their abilities to consume deceased bodies, thus preventing the world from becoming awash in corpses, elevating the lowly maggot to the status of a noble creature indeed (admit it, it would be unpleasant to be awash in corpses). Linnaeus should thus be credited with realizing the importance of the holistic environment (the total assemblage of organisms as well as climate). He clearly saw that organisms interact in nature and he grasped the essential nature of the ecological context in which organisms function. But he also believed that God is good and God made all of nature. That view remained firm.

The teleological view of nature as a balanced, purposeful system under the authorship and stewardship of an intelligent Creator was further promoted by the articulate British writer William Paley (1743–1805).[14] His best-known work is *Natural Theology*, first published in 1802. In this book Paley makes an oft-quoted analogy with a watch. Should you be wandering across a field and come upon a watch, what would you take it to be? Would you assume other than that it was created by design, assembled with parts by an intelligent designer for a specific purpose? Of course you would. Paley's analogy was clear: nothing so complex as a watch could be invented other than by an inventor; likewise the design of nature, from the complexity of a vertebrate eye, to the numerous adaptations of animals and plants, to the assumed great balance of nature, could not have come about other than by intelligent, purposeful design. Given such purpose, Paley further argued that the works of nature could be used to help understand the intentions of the Creator. In other words, Paley fundamentally recycled Francis Bacon's (1561–1626) earlier assertion that to know God, one should know nature.

There was nothing new, unique, or scientific about natural theology. It was then, as it is now, an intellectual blind alley. It merely argued that any interaction in nature, any behavior, no matter how violent or brutal or elegant or whatever, was ultimately the action of an omnipotent and omniscient Creator whose profound wisdom far transcends any meager human ability to account for why nature is as it is. Such a view is, at a certain level, perhaps satisfying, but it remains stunningly vacuous. You learn very little about nature when you assume, a priori, such learning to be ultimately beyond your full understanding. Paley's arguments were for the most part uncritically accepted, as they accorded well with the prevalent societal view that all organisms were, indeed, the products of intelligent design by special creation.

Cultural inertia is hard to alter. It should be obvious that today's push to include intelligent design in biology courses both in high school and college, as a way of "teaching the controversy," is no more intellectually robust than Paley's superficial arguments were in the pre-Darwinian era. It is worse, in fact, because those who promote it should know better. And, as has been borne out in the courts repeatedly, intelligent design, with all of its pleadings about "irreducible complexity," is not a form of science.

One scientist who did not accept Paley's natural theology was Jean-Baptiste de Lamarck (1744–1829), the French zoologist and botanist.[15] Lamarck gets a bad rap. He is often chastised in biology books for his mistaken belief, first published in 1809 in his book *Zoological Philosophy*, in what is called the "inheritance of acquired characteristics." Lamarck's concept was that changes in the form of the body over the lifetime of an organism (such as stronger limbs, for example) are somehow genetically transmitted to the next generation. This would mean, of course, that someone who worked out regularly and thus has well-developed biceps and who also plays Mozart flawlessly on the violin would give rise to muscular children who are talented violinists. Most parents know this does not normally happen.

As incredible as inheritance of acquired characteristics now seems, it took until nearly the end of the nineteenth century before Lamarck's suggested mechanism for evolution was shown to be wrong. Even then, others attempted to "demonstrate" inheritance of acquired characteristics well into the twentieth century.

Even Charles Darwin, who strongly opposed Lamarck's views on evolution, nonetheless answered the siren call of inheritance of acquired characteristics. Only he called it pangenesis. The idea was that tiny "gemmules," undefined and not demonstrated to exist, would somehow find their way to the gonads from the biceps or whatever and then make the necessary adjustments to bring the desired trait to the next generation. Darwin, who had no knowledge of what would become Mendelian genetics, knew his idea was likely incorrect and perhaps regretted even suggesting it.[16]

With careful experiments, August Weismann, perhaps the most influential biologist of the latter part of the nineteenth century,[17] demonstrated that the somatic cells, the cells of the body, are not able to directly or indirectly communicate information to the germ cells, the egg and sperm, that carry the actual hereditary information. The appearance of an organism, its phenotype, is determined both by genetic and environmental influences. The germ cells represent only part of the phenotype, the assemblage of genetic material (called the genotype). Thus some changes in the phenotype, such as a fondness for Mozart, are not directly under the influence of the genotype. Acquired characteristics are not inherited, no matter how much we might wish them to be.

But realize this: Lamarck was the first important, influential evolutionist. He argued unambiguously that populations of organisms could and did change with time into entirely new species. Lamarck's argument for species change was important from an ecological standpoint because it relied heavily on relating the organism, be it plant or animal, to the environment. Lamarck fully accepted the widely held view that organisms are elegantly adapted to various environments. But he also realized that environments typically change given sufficient time. If an organism is already ideally adapted to a given environment, and its environment changes, such a change, Lamarck reasoned, must reduce the adaptiveness of the organism. Lamarck made the straightforward deduction that organisms must either change as well or become extinct. Note the potential impact of Lamarck's reasoning on the balance of nature concept. If environments, both abiotic and biotic, eventually tend to deteriorate around previously well-adapted organisms, this alone argues against a balance of nature. Whatever equilibrium exists must be constantly reinvented. Some balance that is.

Evolutionary biologists have modernized one aspect of Lamarck's view in what they call the Red Queen metaphor or evolutionary "arms race."[18] In short, one reason environments tend to change (thus altering, or more accurately, preventing a real balance of nature) is due to the evolution of the organisms themselves. Recall that the fictional Queen in Lewis Carroll's *Alice Through the Looking Glass* spoke of always having to move in order to stay in place. That is, metaphorically, what evolution is. Movement is measured in evolving adaptations, failure to move in extinctions. If you examine the mammalian fossil record from the onset of the Cenozoic era (one of many examples I could cite), you will quickly see that prey animals, throughout the period of 65 million years ago to the present, have tended to evolve numerous anatomical adaptations (such as longer metatarsal bones allowing greater stride length) that enhance predator detection and avoidance, just as predators have tended to evolve adaptations enhancing prey detection and capture. Cheetahs would have an easy time of it had they lived in the late Eocene, for example. They could have caught anything that moved. At the same time as this arms race progressed, numerous extinctions occurred as well.

Rejecting outright extinction, Lamarck relied on a form of vitalism, and suggested instead that organisms possess an internal vital "force," a kind of "genetic drive" unique to life itself that enables them to change in ways that help them compensate for changes in their environment. These newly adaptive acquired characteristics are then somehow transferred to future generations. Accompanying this notion was Lamarck's belief in improvement. He argued that the vitalistic evolutionary drive inherent in nature ultimately improved the species, made it somehow better. Not only was the balance of nature restored, it was improved! Lamarck suggested that humans evolved from upwardly mobile apes, citing the orangutan as a hypothetical ancestor.

Most scientists, including Charles Darwin, eventually rejected Lamarck's vitalistic view of evolutionary change but not the idea of evolution itself. Give Lamarck credit. He clearly saw that evolution must occur, that only evolution could explain the patterns so evident in nature. He used ecology to bolster that view. But he had no clue about genetics or genes (no one else did, either). So yes, he made up a concept that seemed logical to him but was flat-out wrong. And worse yet (though most critics seem not to focus on this point), his theory for evolutionary change was in no way testable. One simply cannot measure vitalistic traits.

Nonetheless, Lamarck, as a bold thinker, challenged a paradigm. But it was a different sort of observer who made a huge contribution to what would become the science of ecology, and he did his work even before Lamarck. His observational skills in the area of natural history, coupled with the groundbreaking work of Lamarck (in arguing forcefully for the reality of evolution), would really provide the prerequisites to what Darwin would accomplish.

In July of 1985, I was doing some writing in Oxford, England and decided to make a pilgrimage. I traveled to the village of Selborne, there to visit "The Wakes," the name given to the home of someone who, in my view, was the first real ecologist. That person was the Reverend Gilbert White, who resided at The Wakes for most of his 73 years. I went there to pay homage to this unique figure in ecological history.

Gilbert White was born in 1720, nearly 100 years before Darwin. Educated at Oxford, White was ordained in 1747. In addition to his duties on behalf of his parishioners in Selborne, White was stunningly adept at the study of natural history. In an age when most insights into nature were gleaned from specimens returned from long sea voyages of discovery, White proved that careful observation within one's own backyard could be equally, if not more, insightful.

White's observations were published in 1789, just a few year's before his death in 1793, in what has become one of the most classic books of its kind, *The Natural History of Selborne*. The book consists of a series of letters, all originally sent to either of two friends, then subsequently revised and edited for publication. The letters take the reader through the seasons at Selborne, exploring the behavior of creatures ranging from insects to frogs, tortoises, birds, and mammals. The letters fascinate, populated as they are with sophisticated observations and speculations. White also loved his subject matter. His deep interest in the phenomenon of hibernation led to a long study of a tortoise named Timothy, an animal originally belonging to his aunt. White cared for Timothy for many years as he studied the animal's behavior, and Timothy's shell is now on permanent exhibit at the British Museum of Natural History.

What was most impressive about Gilbert White was how meticulous he was in his observation of nature. What an empiricist! At a time when many naturalists based their writing on either their own musings or the reputed discoveries of others, White wrote only from first-hand experience in the field. For example, he was the first to discover that a small, nondescript bird then commonly called the willow-wren was not one bird species at all but three! Today ornithologists know these birds as the wood warbler (*Phylloscopus sibilatrix*), the willow warbler (*P. trochilus*),

and the chiffchaff (*P. collybita*). To the casual eye, these three warbler species look virtually identical: all three share the same body
size, bill characteristics, and overall plumage coloration. All three
feed actively high in the foliage, where they capture small arthropods. But closer inspection reveals that the wood warbler is more
brightly colored, a more intense yellow-green than the other two;
that the willow warbler looks plumper than the chiffchaff; that the
chiffchaff is more gray than either of the other two. White noticed
these subtle differences and more. He noted that the birds' songs
are each distinct. Though the three species look alike and may occur
within the same area, they do not interbreed because their unique
songs enable species recognition. Ecologists are now familiar with
many examples of species that look similar but that are ecologically
segregated and reproductively isolated, one species from another,
either by plumage, song, habitat preference, or some combination
of those factors. But it was Gilbert White, about 250 years ago,
who first noticed it. Gilbert White exemplified how to apply an empirically data-based approach to natural history, indeed, how to do
ecology.[19]

Other areas of the natural sciences flourished. Studies in geology, anatomy, embryology, and biogeography all provided evidence for an underlying unity among the many kinds of organisms
on Earth. The concept of a balanced nature was not challenged by
such insights. Rather, it was strengthened.

Needed was a fresh view and a grand synthesis of this increasingly large base of information, a sound, logical theory documenting not only that evolution does occur but also how it occurs. That
task of discovery would fall to Charles Robert Darwin and another, younger man, Alfred Russel Wallace.

5

Ecology A.D. ("After Darwin")

Charles Robert Darwin and Abraham Lincoln were both born on February 12, 1809. Lincoln was assassinated on April 14, 1865. Darwin would survive until April 19, 1882. It is hard to overstate the importance of either man. Each was an agent for profound change.[1]

Charles Darwin, with stimulus from a younger man, Alfred Russel Wallace (1823–1913),[2] solved what is arguably the greatest problem in biology. Darwin, borrowing an expression first used by the astronomer John Herschel, called it "that mystery of mysteries," the means by which species form. Darwin and Wallace discovered a mechanism for evolutionary change. Natural selection is a blind and mechanistic process lacking any vitalistic component (and thus it is not Lamarckian). It is non-teleological, with no underlying purpose. It is a blind process that inevitably follows from two realities: (1) the existence of genetic diversity conferring different phenotypic traits on individuals within a population, (2) and the inevitable limitation of resources. When resources are limited, those individuals with traits most suitable (i.e., adapted) to the environment will have a statistically better chance of survival and reproduction. Survivorship is thus not random, but dependent on characteristics of phenotypic variants within the population. Natural selection would seem to be as much a characteristic of the universe as gravity and electromagnetic radiation. If a planet somewhere, anywhere in the universe has life, natural selection will be blindly acting on the various life forms.

Darwin and Wallace each independently discovered the theory of natural selection. Documentation of Darwin's discovery dates to notebooks on species transmutation that he kept in the late 1830s, but it was not until 1842 that Darwin first dared write a brief sketch outlining his theory. After this first tentative description of natural selection, he followed with a much more detailed essay in 1844. He took pains to be sure his wife Emma (who never accepted her husband's views on evolution) would see to its publication should he die before completing his major work on natural selection and evolution.

In 1855, prior to penning his essay on natural selection, Wallace wrote an insightful essay arguing that all species arise from previously existing species. Wallace, by then convinced of the truth of evolution, knew he needed some mechanism to account for species change. That was solved by natural selection, an insight that came to Wallace in 1858 while recuperating from illness in the Malay Archipelago. Knowing Darwin's eminence as a naturalist, Wallace mailed his essay to Darwin asking him to see to its publication. Thus, at minimum, Darwin had written (though had not published) on natural selection fully sixteen years prior to Wallace. Darwin's belief in evolution dates back a bit earlier, to 1836.

Scholars debate Wallace's role in this profound revelation of how biology actually works. Who really deserves credit for discovering *the* paradigm? Some suggest Darwin was unfair, waiting too long, getting scooped, and then grabbing the lion's share of credit, when, in reality, Wallace really should have had priority. Darwin is accused of plotting with his close friends Charles Lyell (an eminent geologist) and Joseph Hooker (an eminent botanist) to assure his priority in formulating natural selection theory, duping Wallace into accepting his secondary role. In my opinion, that is simply wrong.

Wallace sent his natural selection essay to Darwin, hence Darwin's dilemma as to how to deal with it. Reading Wallace's essay must certainly have been a shock to Darwin, and he was understandably dismayed. But he had shared his natural selection theory with both Lyell and Hooker years earlier, and he sought their counsel, as he wanted to treat Wallace fairly. At the urging of Lyell and Hooker, essays from each man were read at the prestigious

Linnaean Society in London on July 1, 1858. Interestingly, those readings did not cause much of a stir among the scientific elite. It was only when Darwin's book was published sixteen months later that the would-be new paradigm was really noticed. Darwin has always received most of the credit (or damnation, depending on whose opinion is aired) for the work, the reason why the term "Darwinism" is still used today. It is doubtful that Wallace could have stimulated the degree of attention that followed Darwin's book, as Wallace had not devoted years to collecting evidence to support his views, as Darwin had.

Credit Wallace with stirring Darwin, long reluctant to go public with his theory, into action, into finally publishing a theory he had pondered for over two decades. And that was Wallace's real contribution.[3]

Why was Darwin so reluctant to publish until Wallace forced the issue? He likely understood the profound impact his theory would have and could not convince himself to present it until he satisfied himself that he had overwhelming evidence to support it. Darwin undoubtedly knew a mechanistic theory would be far harder to accept than one that still left room for some sort of vitalistic divinely imposed direction. Beyond that, in 1844, the very year Darwin wrote his long essay on natural selection (which was to become the basic outline for the *Origin*), a book with the Victorian title *Vestiges of the Natural History of Creation* was published.[4] The author, Robert Chambers, published it anonymously, as he knew only too well what a commotion it would cause. And it did. Chambers presented a sweeping overview of evolution, ranging from the universe to humanity. It was a complex book and nowhere nearly as tightly or coherently argued as what Darwin would subsequently publish. It was largely teleological but it did advocate organic evolution. The strongly negative comments it engendered from the scientific elite must have been sobering to Darwin. Society did not seem ready to take evolution either seriously or graciously, and Darwin likely knew his coldly mechanistic theory, if anything, would be perceived as even more inflammatory.

The *Origin* was the culmination of Darwin's thinking that almost certainly began while Darwin was still riding the waves (and being seasick) aboard the *Beagle*. His many observations, particularly on the curious distribution of island flora and fauna (especially as observed on the Galápagos Islands), planted the first seeds of doubt about the immutability of species. Upon his return to England in 1836, Darwin began his long intellectual journey, the first leg of which would culminate with the publication of the *Origin* on November 24, 1859.[5] Darwin would then go on to write other books, each of which would focus on some aspect of evolutionary theory. Though Darwin only hinted at human evolution and common descent with apes in the *Origin* (the first of the Neanderthal skeletons was discovered in 1856), he would deal in detail fully with the origin of humanity in *The Descent of Man and Selection in Relation to Sex* (1871), followed by *The Expression of Emotion in Man and Animals* (1872). His final book was *The Formation of Vegetable Mould, through the Action of Worms* (1881). In the worm book Darwin argued for how small but continuous forces (the action of earthworms in soil), over long time periods, could result in dramatic changes in landscapes (such as toppling monoliths at Stonehenge). That is a forceful analogy for how natural selection results in major evolutionary changes through small cumulative changes over time.

The *Origin*, which Darwin referred to as "one long argument," made two profoundly important points. The first was that all organisms extant on Earth, as well as all extinct forms that preceded them, ultimately share a common ancestor. In Darwin's view, all life forms are related through a profound genealogy that extends back to the time of the first appearance of life on the planet. Life's history is thus visualized as a dense genetic bush of variously related forms (a pattern commonly called "the tree of life"). Life forms change, evolve with time, many become extinct, but each shares a historical genetic connection with all others. Extinction represents dead stalk tips on the bush. Speciation represents newly branching stalks. Darwin used the metaphor of a many-branched bush to illustrate this concept, which he repeatedly called "descent with modification."

The second part of Darwin's "argument" was to propose a mechanism for how evolution occurs, called natural selection. But what is natural selection? What is the process that inspired Darwin's friend and defender, Thomas Henry Huxley,[6] to declare "How stupid of me not to have thought of that"?

Selection, by its very name, implies change. If certain traits are selected, others will in effect be rejected. If, for instance, diners in a restaurant never order the halibut but often order steak, the menu will soon change such that halibut is no longer included, while several kinds of steak are featured. If traits are genetically based, then gene (or, more accurately, allelic) frequencies change as selection acts.

The most familiar examples of biological selection are seen with domestication of plants and animals, a point not lost on Darwin. He used domestication as an analogy to begin the *Origin*. Charles Darwin was deeply involved with breeding exotic pigeons and commented at length in the *Origin* about how modern pigeon breeds, the so-called pouters, tumblers, jacobins, fantails, and nuns, were all direct descendents of the wild rock pigeon (*Columba livia*).[7] He made essentially the same comment about how some modern dog breeds, as different as they are, may trace their lineage back to wolves and foxes. The point of the analogy was to convince his readers that there is hidden within each species the strong genetic potential to change with selection. Domestication demonstrates that fact with crystal clarity.

Natural selection is the process responsible for adaptation. The phenotype of any organism, most of what you see, results from the cumulative effects of natural selection. Selection acts on anatomy, physiology, and, in animals, behavior. It is the force in nature that quite literally shapes all living things. It explains why African lions (*Panthera leo*) have such long canine teeth and powerful jaws; why certain wasps paralyze spiders and then lay their eggs on them; why red mangrove trees (*Rhizophora mangle*) have stilted roots; why some flowers are tubular in shape and red in color; why some molds produce chemicals that inhibit bacterial growth. Aristotle would be amazed if he only knew. But how does natural selection actually work?

The key to natural selection is Malthusian economics. Adam Smith (1723–90), the great advocate for unrestrained capitalism (including his analogy of "the invisible hand" acting on economic systems), provided the intellectual model for natural selection, the model of free, unrestrained competition (among corporations) for limited resources (the marketplace). Malthus applied Smith's views to human social policy and arguments about Malthusian economics have raged throughout the halls of academe ever since.

Both Darwin and Wallace read Thomas Malthus's *An Essay on the Principle of Population*.[8] This work, first published in 1798, went through six editions, the last one published in 1826, when Darwin was seventeen years old. The crux of the essay for both Darwin and Wallace was this: Malthus described how basic environmental resources would ultimately limit populations, thus creating a struggle for existence. Competition is inevitable (because of population growth) and thus only some within a population ultimately survive, while others must perish. What Darwin and Wallace did was apply the Malthusian principle to nature, with one key addition. They realized that individuals vary genetically within populations. This variation will ultimately influence who survives when it is crunch time. Those variants most suited to whatever the environment imposes will tend to survive and reproduce better that those with different traits.

Note please that the "selection" that occurs is metaphoric and not guided. Selection is a statistical truth, an inevitable result from circumstances facing populations. I will say more on this point below.

Unlike Newton, who generated a precise mathematical formula for predicting the gravitational attraction between bodies, Darwin and Wallace articulated their theory of natural selection in words. It was a scientific, empirically testable theory expressed in pure logic. Darwin and Wallace both realized that there is a Malthusian-like "struggle for existence" throughout nature. The intensity of the struggle varies depending upon many factors. If resources are ample and population density is small, there are probably sufficient resources for all members of the population. There will be little or no struggle for existence and thus selection is "relaxed." But when any essential resource becomes limited, there will inevitably be some form of competition, direct or indirect, within the population for access to that particular resource. Not all will obtain it, and, if it is vital to survival or reproduction, not all will survive or reproduce.

The struggle for existence need not be obvious, as it can also be against the elements of nature, factors such as cold temperature, ice storms, protracted drought, or fires. If the climate should cool, for example, such an event would place a stress on organisms adapted to a warmer environment. Selection also results from effects of predators, parasites, and pathogens. Not all individuals will be equally skilled at avoiding predators or equally resistant to the ravages of parasites or pathogens. Darwin wrote, "I should premise that I use the term Struggle for Existence in a large and metaphorical sense."[9]

Natural selection shows clearly that survival and success in reproduction are not random in nature. This point is exactly the opposite of what many people believe about natural selection and why they don't understand it. It is common to hear that "we didn't get here through a random process." Right, we didn't. Individuals in a population are not alike, but differ from one another in significant ways. These differences are largely genetically based and they become of monumental importance when the individuals in a population are, indeed, engaged in a struggle for existence. Those individuals most suited to the conditions imposed by the environment will, logically enough, be most likely to survive and subsequently reproduce. In evolutionary terms, such individuals are the most fit.

Fitness is the relative reproductive success of an organism, dependent on how it fares under the present environmental conditions compared with others of its population. If the environment chills, those within a population with the thickest fur, or warmest down, or who can secure the snuggest den, are most likely to survive and reproduce. They are on average more fit than those with thinner fur. This concept, the essence of natural selection, is often called the "survival of the fittest," though it was not Darwin but Herbert Spencer who coined the term that has become the catch phrase of Darwinism.

Many are born, but not all survive to reproduce. Those that do are a select, nonrandom subset of the original cohort. Because

individuals differ genetically, the population evolves as environments change in various ways. As noted above, Darwin recognized
that fitness is largely a metaphorical term and need not literally
mean "nature red in tooth and claw." Organisms may struggle
against each other, against other species, or against the elements.
But ultimately it will be the genetic endowment of the various individuals that will weigh most heavily in the struggle for existence
and the ultimate survival of the fittest.

Natural selection could not occur without genetic variability. It
is the genetic variants that are sorted, a sorting that can take any
direction, depending on circumstances within the environment and,
indeed, generated by the environment, both living and nonliving.
Many factors provide genetic variability upon which selection acts:
mutation, gene flow among populations, genetic recombination during sexual reproduction. And it is here that there is randomness,
since mutation, the ultimate source of all genetic variability, is largely
a random process.

Natural selection is not prescient. It has no way of looking into
the future. Today's well-adapted creature may be extinct tomor-
row. There is thus no innate directionality to natural selection.
Natural selection acts only for the moment and does not "plan"
for the long-term future. It is a blind sorting process. To measure
natural selection in nature it is necessary to show what the agent
of selection is and what environmental factor is responsible for the
differential survival or reproduction within a population.

It required almost a century after Darwin and Wallace described
it before natural selection was shown to actually occur in nature, in
the famous example of the British peppered moth (*Biston betularia*),
now described in virtually every introductory biology text as "industrial melanism." As tree bark became soot-covered by industrial
pollution, the normally light-colored moths fell prey to birds but
the rare darker moths, normally less fit, survived better as they were
better camouflaged (thanks to the soot covering tree bark). The
moth population gradually evolved from light to dark. Many examples of natural selection in nature have been described since.[10]

A clear and dramatic example of natural selection was documented for the medium ground-finch (*Geospiza fortis*) of the Ga-

lápagos Islands.[11] The Galápagos Archipelago is 600 miles west of Ecuador, in the Pacific Ocean. It is subject to climatic fluctuations such as periodic droughts and occasional El Niños, when much rain falls on the islands. In 1977 there was an extremely severe drought throughout the islands.

The finches were studied on the small island of Daphne Major, where it was possible to capture and band each individual and actually follow its fate. During the drought, most of the plants on Daphne Major failed to produce seeds, the normal food of the finches. Most of the small-seeded plants were seedless, but a few plants that made large, hard seeds persisted. The medium ground-finch population, now experiencing a severe food shortage (brought on by the drought), was devastated, decreasing by 85% from June of 1976 to January of 1978, mostly from starvation. But some survived. Those that survived had the largest, deepest bills of any of the birds and were thus able to crack the hard seeds, the only resource left to them during the drought. The difference in bill size between survivors and the many that perished was a mere 0.5 mm, a difference undetectable to observers but measurable (with calipers) and critical in its importance. Again I emphasize that survival among the finches was not random: only those with large bills survived. Bill size is genetically based, and the offspring of the survivors had similarly large bills. Strong evolutionary change was occurring in as little as a single generation, due to the high intensity of natural selection caused by the drought.

In 1982–83 there was a severe El Niño throughout the Galápagos and much rain descended on the islands. This created an abundance of food resources for the finches as the islands turned rich green and seeds abounded. But small-seeded plants produced many more seeds than large-seeded plants and, again, bill size shifted among the medium ground-finches. Those with smaller bills acquired more food and left more offspring. Bill size moved back toward what it had been before the drought of 1977. Such is the nature of natural selection. It responds only to the situation at hand. There is no "good gene" or "bad gene" in any absolute sense. It all depends on resource availability, on the conditions at the time of the selection event.

Finally, note that selection may work in any of three ways. One common and often overlooked way is that of *stabilizing selection*. If an albino squirrel is born it is not likely to survive detection by predators as well as a normal colored squirrel. Thus populations maintain a narrow phenotypic range based on selection on either end of a bell-shaped curve. Extremes are eliminated.

The form of selection most important for evolutionary change is *directional selection*. That happens when some aspect of the environment changes, imposing novel selection pressures. Think of it as acting on one end of the bell-shaped curve, pulling the curve toward a different phenotypic mean. The evolution of larger bill size in the drought-stricken Darwin's finches is such an example.

Disruptive selection occurs when selection acts on parts of the phenotypic range, breaking up the bell-shaped curve into several smaller, narrower phenotypic ranges. Disruptive selection may be important in certain patterns of speciation and mimicry.

Natural selection is conceptually easy but philosophically difficult to understand. Some viscerally dislike it for its lack of teleologic purpose or direction toward "improvement." To others it seems an incredible game of chance. But it isn't: organisms do not evolve "randomly" with regard to adaptations essential for survival. Quite the opposite is the case. Anyone who says that natural selection is a random process is flat-out wrong. As stated above, only mutation, the ultimate source of genetic variability, is random.

Another difficulty with selection was what almost killed it by the close of the nineteenth century. Because biologists had no firm understanding of what would become known as Mendelian genetics[12] and, later, the field of population biology, it seemed to them that natural selection was simply too weak to actually work. The analogy was offered of a drop of black paint in a can of white paint. Black may be beneficial, but it would merely be "swamped" as it "blended" in a sea of white paint. So how could rare though beneficial traits actually increase?

That problem was firmly solved in the 1930s when Ronald Fisher, J.B.S. Haldane, and Sewall Wright, each acting independently, were able to refine the theory of natural selection, coupling it with Mendelian genetics and Hardy-Weinberg population genetics. These and other evolutionary biologists were thus able to effectively utilize mathematics to predict just how gene frequencies change under different kinds of natural selection. The work of Fisher, Haldane, and Wright soon came to be known as neo-Darwinism, or the "new synthesis." It represented a mid-twentieth-century rebirth of evolutionary thought.

The *Origin* is insightful to any ecologist. I would argue that you could call it the first decent ecology text. The very basis for natural selection, the tendency of any population to increase exponentially, is fundamental to the study of population biology. Many other contemporary ideas in ecology appear, often in somewhat rough or vague form, in Darwin's writing. He provides elegant examples of how ecosystems radically change as a result of such things as grazing, for example. He speaks of intra- and interspecific competition and the effects that such interactions might have on the evolution of organisms and speciation. He writes of how seeds are dispersed on mud that is stuck to the feet of far-flying waterfowl. He muses about why there are so many more kinds of plants and animals in tropical latitudes than temperate or polar regions. He describes how different kinds of insects are uniquely structured to pollinate specific plants, an example of what ecologists now call coevolution. Darwin's final paragraph in the *Origin* begins with an unabashedly ecological example:

> It is interesting to contemplate an entangled bank, clothed with many plants of many kinds, with birds singing on the bushes, with various insects flitting about, and with worms crawling through the damp earth, and to reflect that these elaborately constructed forms, so different from each other, and dependent on each other in so complex a manner, have all been produced by laws acting around us.[13]

Note that there is an implicit assumption of balance in this paragraph. When Darwin writes of organisms dependent on each other in so complex a manner, he likely thought natural selection accounted for what he still believed to be a balance of nature.

It is difficult to overstate the importance of Darwin's work.[14] Whether at the time one agreed or disagreed with his theory that

evolutionary change and adaptation were brought about principally by natural selection (and most, including his ardent support-
ers, disagreed or had reservations), Darwin convinced most readers of the reality of evolution itself. His work clearly accelerated
interest in natural history and provided the foundation for a research program that eventually would blossom into the study of
ecology. Recall that it was only seven years after the publication of
the *Origin*, 1866, that Haeckel invented the word "ecology" to describe the study of the "economy of nature," a term Darwin used
four times in the *Origin*.

Darwin argued forcefully against creationism but was far more
accepting of the balance of nature. He provided an oft-cited example in the *Origin of Species* that demonstrates a "balance" within
a food chain. He observed that clover requires pollination specifically by humble-bees (which are the same as bumble bees), a kind
of bee that nests in the ground. Humble-bee nests are preyed upon
by field mice. Therefore a large number of mice would reduce the
population of pollinators and thus render the clover less able to
reproduce. Predators of mice, which, themselves, have nothing to
do with clover, are thus essential to the health of clover populations. Without the proper "balance" of stoats (weasels) and cats to
kill mice, the elegant pollination system upon which clover depends would come crashing down.[15]

Darwin's perhaps subconscious assumptions about nature's balance were hard to reconcile with his view of natural selection acting opportunistically, adapting populations to changes in their environments. This led to one of his most famous metaphors, the
wedge. Darwin assumed, because he was educated in the paradigms of Western civilization, that nature was balanced and thus
all of the available space in the economy of nature must be filled.
Darwin believed in the philosophical concept of plenitude, a full-
ness or completeness to nature. There was really no reason why he
should think this other than that it was a philosophical bias he assumed to be true rather than critically analyzing it. Thus he was
puzzled by how a newly evolved population could enter an otherwise full habitat. He satisfied himself by arguing that "The face of
Nature may be compared to a yielding surface, with ten thousand

sharp wedges packed close together and driven inwards by incessant blows, sometimes one wedge being struck, and then another with greater force."[16]

Newly evolved species thus entered by bashing their metaphorical way into the ecosystem, becoming part of the balanced economy of nature. Elsewhere *Origin* in Darwin asserted,

> Battle within battle must ever be recurring with varying success; and yet in the long-run the forces are *so nicely balanced*, that the face of nature remains uniform for long periods of time, though assuredly the merest trifle would often give the victory to one organic being over another.[17]

The balance of nature was clearly assumed and included in Darwin's otherwise radical book. Thus it is little surprise that the natural historians who followed him in the nineteenth century clung to the notion of balance. Two in particular are important to the development of ecology as a discipline: Karl Möbius and Stephen A. Forbes.[18]

Karl Möbius (1825–1908) studied coastal marine ecosystems. He was impressed by how many different species of organisms cohabit oyster banks. In 1877 Möbius published a paper with the title "An Oyster-Bank Is a Biocönose, or a Social Community." In it he described how the mere presence of oysters permits the existence of a widely varied community of arthropods, mollusks, echinoderms, and coelenterates, as well as various kinds of algae and other plants. Möbius's work helped define what would become the concept of the ecological community, a "balanced community," and his focus on the unique importance of the oyster was an early description of what we now call a keystone species (chapter 12).

In 1887 Stephen A. Forbes (1844–1930) published a paper that soon became one of the most influential in the emerging field of ecology.[19] In "The Lake as a Microcosm," Forbes linked the interdependencies among the various living organisms of a lake with their collective dependency on the nonliving environment, ultimately viewing the lake as an example of the balance of nature. Forbes wrote that "Perhaps no phenomenon of life in such a situation [within a lake] is more remarkable than the *steady balance of organic nature* [italics mine], which holds each species within the limits of a uniform average number, year after year, although each one is always doing its best to break across the boundaries on every side."[20]

Forbes suggested that ecosystems develop in such a way as to eventually attain order from chaos, a theme that would occupy ecologists throughout the next century as they tried to make the concept of balance of nature into a scientific rather than metaphorical paradigm.

Ecology after Darwin was not the same. Natural history was no longer natural theology. It had become science, the science that studies the balance of nature. Ecology still had a long journey ahead of it.

The Twentieth Century
Ecology Comes of Age

harles Darwin got ecology launched. He described what really happens in the economy of nature far better than anyone who preceded him. Ecologists then promptly forgot about him for something like a half a century. Even then they were a bit slow on the uptake. Ecologists were conspicuously absent from the grand synthesis of evolutionary theory that was started in the 1930s with the seminal work of J.B.S. Haldane, Ronald Fisher, and Sewall Wright. When Theodosius Dobjhansky and Ernst Mayr developed the biological species concept,[1] the notion that reproductive isolating mechanisms separate species, ecologists contributed little. George Gaylord Simpson's classic work interpreting the fossil record for trend and rate analysis,[2] at least to my knowledge, never found its way into ecology texts, though it certainly had significant ecological implications.

For the early years of the twentieth century ecology remained essentially a descriptive science. Ecologists went into the field, counted plants and animals, made lists, and that was pretty much that. To give them their due, the natural world is complex and ecologists found it challenging enough to accurately survey the species present in various environments and obtain estimates of population sizes.

The emergence of lab-based physiology served as an example to would-be ecologists, who tried, in their field methodology, to emulate the methodology of laboratory physiologists. Ecologists were trying hard to shed the image of bug collectors and attain the respect accorded laboratory scientists, the practitioners of the "scientific method."

Two subdisciplines of ecology were quick to materialize, plant ecology and animal ecology. This was hardly surprising given that botany and zoology had, by then, emerged as concentrations within academe. Of course the natural world is more diverse, as it includes many kinds of microbes and protozoa as well as perhaps millions of species of fungi. But tools for studying these organisms did not exist as they do now, and ecologists found plenty to do in the study of multicellular plants and animals.

The Ecological Society of America was founded in 1915, with Victor E. Shelford (1877–1968) as its first president.[3] A grand total of sixteen people joined. Today the membership is nearly 10,000, and each year about 3,000, mostly eager graduate students, attend the ESA's annual meeting.

Shelford pioneered the study of animal ecology, and his research ranged widely. He worked on describing food webs and was among the first to perform serious research on problems of water pollution. With America's premier plant ecologist, Frederic Clements, Shelford coauthored *Bio-Ecology*, published in 1939, one of the first comprehensive ecology textbooks.

In 1963, as one of North America's most senior and distinguished ecologists, Shelford published *The Ecology of North America*, a monumental work that describes most of the major terrestrial ecosystems on the continent.[4] The level of detail contained in this book is staggering. Shelford attempted to summarize the plants and animals as well as the climate and soils that typify ecosystems ranging from tundra to the subtropics of Florida. For example, Shelford cites a study done on May 4, 1935, at Turkey Run State Park, Indiana, concluding that there are 10,000 red-backed salamanders per hectare and 3,868,004 invertebrates (mostly insects) per hectare. These kinds of descriptive studies were the grist for ecology in its early years.

Shelford's contribution to conservation was his leadership in chairing the Committee for the Preservation of Natural Conditions in 1917 and forming a group called the Ecologists' Union in 1946.

This became the Nature Conservancy in 1950.[5] One of Shelford's students was Eugene P. Odum, who became a leading figure in the development of ecosystem ecology and authored the first truly widely used ecology text, *Fundamentals of Ecology* (1953).

Descriptive ecology gradually was augmented by an attempt to look for and understand patterns in nature. Several directions emerged. One was to understand how communities, especially plant communities, are organized. Another was to learn how energy moved from one component of an ecological community to another. A third was an attempt to understand what regulated population density, the Malthusian parameter that is so essential as a prerequisite for natural selection. Note that each of these three areas assumed that nature achieved and then maintained equilibrium, a "balanced" condition. Ecologists clearly recognized that strong interdependencies and subtle interactions are the very fiber of nature. It was as though a dense Gordian knot was before them, needing to be unraveled.

Plant ecologists devoted much energy to learning how ecological communities develop with time, and thus the study of ecological succession dominated the agenda. Researchers who described themselves as "phytosociologists" attempted to account for why certain aggregations of plant species co-occur, an effort to understand how species in plant communities are integrated into a presumably balanced assemblage.[6]

In North America, one school of plant ecologists, championed by Frederic E. Clements (1874–1945),[7] thought the ecological community to be a kind of "supra-organism" that went through various (and often complex) predictable stages toward maturity, eventually forming a "climax community." An assumption that communities reach a balanced equilibrium was deeply entrenched in this view.

Clements's perception of the community was largely derived from his holistic (read "balance of nature") philosophy coupled with his detailed studies of ecological plant succession (often called "old field succession," as it takes place typically after agricultural abandonment of land). An area of bare soil devoid of any organisms is typically colonized first by rapidly growing herbaceous plants, such as various forbs (e.g., goldenrods and various other "weeds") and grasses. Eventually woody shrubs and trees invade, each invasive wave constituting a stage of development analogous, in Clements's view, to growth stages in an organism. Clements called these developmental stages "seral stages," and the entire process of community development constituted what he called a "sere." Eventually the "climax" of the process would be attained, the growth stage at which the community became self-reproducing. Clements viewed the climax as ultimately determined by regional climate, and, for that reason, termed it the "climatic climax."

Clements believed that interspecific competition among plant species largely drove the succession process, resulting eventually in a stable (again, read "balanced") association of coexisting species. He recognized that there were numerous developmental trajectories by which a community of plants within a given region could gradually change. In other words, although he thought that each community "should" develop into the regional climatic climax, he saw clearly that many did not and he had to account for that uncomfortable reality. He coined an extensive and cumbersome jargon in an attempt to accommodate the numerous possible avenues to the climatic climax, often a bit of a tortured pathway. For example, Clements wrote of such things as proclimaxes, subclimaxes, disclimaxes, and postclimaxes. He described associations of plants as well as consociations, faciations, lociations, and societies. Encompassing all of the complex terminology was the assertion that the plant community is real, is integrated, and develops in a predictable, orderly, deterministic manner.

Clements's view did not go unchallenged. Good thing.

One who was not persuaded was Henry A. Gleason (1882–1975).[8] Gleason, like Clements, conducted ecological research in the Midwest, but in a region where prairie and forest both existed, indeed often intermingled. He recognized that forests prevented the invasion of species typical of prairies because the shade of the forest did not permit such species to receive adequate sunlight needed for them to become established. He also recognized that prairie grasses, with dense root systems acting to significantly thicken the sod, prevented forest species such as trees and shrubs from becoming established. Thus Gleason did not envision plant communities as moving toward a regional climatic climax but, instead, recognized that biotic factors could act to maintain a community indefinitely in one place such that it differed considerably from other nearby communities.

Gleason's analyses of plant communities suggested that the assemblages were ultimately dependent first on seed dispersal and then on subsequent establishment of various plant species. The determination of which species would persist was largely dependent on their physiological tolerances in relation to the local conditions as well as stochastic factors pertaining to such life cycle characteristics as dispersal. In Gleason's view, no two ecological communities were sufficiently similar to conclude that they should be viewed as "supra-organisms" in the Clementsian sense.

Gleason termed his view of the ecological community the "individualistic concept of the plant association." The use of the term "individualistic" was meant to indicate that plant communities are sufficiently distinct, one from another, that viewing them as moving intrinsically toward a common climax state is misleading. In contrast to Clements's holistic, "balance of nature" view, Gleason's view was much more stochastic (influenced by chance events). Gleason's concept of the plant community contained a strong focus on the importance of site-to-site variability that was lacking in Clements's conception of the community. In a Gleasonian community, species continually immigrated into the community while other species became extinct from it such that the plant community was an ever-changing tapestry, what Gleason once called a "kaleidoscope."[9]

Gleason's view was clearly not based on any assumption of equilibrium. It anticipated modern thinking on the ecological community. The debate raged through the middle part of the twentieth century, when new techniques in plant community analysis began to strongly favor the Gleasonian individualistic model.

One advance was the application of ordination techniques to study plant communities.[10] This approach is multivariate, measuring what are termed *importance values* of each plant species. To compute an importance value and then apply it to community ordination, it is necessary to measure the species' distribution on several axes (each of which represents a variable such as soil pH) followed by determining which variables best account for the distribution of each species. The ordination approach measures the density per acre, basal area per acre, and relative frequency (the percentage of stands in which the species is found to occur relative to that of all other species) for each plant species. These data are used to compute an importance value for each species. Importance values for each species in each stand are then graphed against a *continuum index*. When the final ordination is determined, it is possible to quantify both the degree of difference among mixed species stands and the precise distribution of each tree species positioned graphically along the continuum index axis.

Ordinations typically resulted in a continuum pattern such that each plant species was uniquely distributed along the continuum. If the plants formed integrated communities this result should not have occurred. Instead, ordination studies clearly suggested individualistic distributions for each species, a result far more consistent with Gleason's concept of the community than with that of Clements.

A variation of the continuum approach was to use an environmental gradient as an independent measure against which to compare the distribution of plant (or animal) species.[11] Robert Whittaker, who pioneered gradient analysis, used gradients to test the concepts that interspecific competition determines community structure and that association of a given species with others in the community is a powerful determinant of community composition. Both of these hypotheses are contained within the Clementsian view of the plant community, and Whittaker showed that neither was supported by the data. The technique was rather simple but effective.

When numerical abundances or importance values of various plant species were ordered along gradients such as slope, or moisture, or pH, each plant species was typically distributed along a bell-shaped curve. What was important was that while these bell-shaped curves overlapped, they were in most cases statistically distinct from each other. In other words, the curves were partially overlapping but not clustered in synchronous groups. There were no patterns that suggested clear segregation between assemblages of species, as would be the case if competitors excluded one another or if groups of species were coevolved to remain together. As with community ordination by continuum analysis, gradient analysis supported Gleason's view that ecological communities, while often sharing numerous species in common, are nonetheless each fundamentally individualistic.

The Clementsian view of the plant community did not withstand careful scrutiny. Gleason's concept of the plant community as an individualistic assemblage fit the data much more closely. However, one must be mindful of the fact that the various tests of the two hypotheses cited above relied only on numerical distribution of plant species. Ecologists still had years ahead before they fully embraced the reality of what an individualistic community really is.

The word "ecosystem," coined by the British ecologist Arthur Tansley (1871–1955), first made it into ecological jargon in 1935 and has been recognized as an organizational level within the discipline ever since. Tansley moved ecology toward a broader-based research program, one that flourished with the work of Charles Elton, who pioneered the study of ecological energetics.

Charles Elton (1900–1991) of Oxford University was a true pioneer of ecology. In his book *Animal Ecology* (1927),[12] Elton made a valiant attempt to construct an ecological flow diagram of what would appear, at first glance, to be a rather simple ecosystem, the Arctic tundra. Elton's diagram, made from his studies on Bear Island (south of Spitsbergen), shows the complex pathways that energy and nutrients follow in moving through the various organisms in the ecosystem. Elton's diagram had organisms arranged in discreet boxes connected by arrows indicating the direction of energy flow. For example, the box "Diptera" (insects such as flies) had an arrow going from it to purple sandpiper and then an arrow from the sandpiper to arctic fox. There was also an arrow from "plants" to the sandpiper. The diagram appears on the same page (58) with another diagram showing the general food relations of the herring to other members of the North Sea plankton community. Both the Spitsbergen and herring diagrams look weblike and complex, a point Elton recognized. The question was how to cut through such complexity to uncover a meaningful pattern. Elton succeeded when he described a pattern that was to form the basis for one of the central tenets of ecology, the Eltonian food chain, also commonly known as the grazing food chain.

Plants take energy directly from the Sun, forming the first link in the Eltonian food chain. Herbivores, such as rabbits or caterpillars, to select but two examples, eat plants. The energy acquired by herbivorous animals is two steps from the Sun. A blue jay that eats a caterpillar is three steps from the Sun, and a Cooper's hawk that eats a blue jay is four steps from the Sun. Elton noted that organisms become generally less abundant and larger in body size the more steps they are along the food chain. Plants vastly outnumber caterpillars and rabbits, which, in turn, clearly outnumber blue jays and hawks. Such a pattern is typical of grazing food chains, though it is sometimes expressed as biomass rather than numbers (i.e., the plants far outweigh the herbivores and the herbivores far outweigh the carnivores). The most accurate expression of the Eltonian pyramid is a pyramid of energy, measured in units such as kilocalories per meter squared per year. Far more Kcal/m^2/yr pass through herbivores than carnivores regardless of numbers or biomass.

Elton wrote of this pattern, "There are, in fact, chains of animals linked together by food, and all dependent in the long run upon plants. We refer to these as 'food-chains,' and to all the food-chains in a community as a 'food-cycle.'"[13]

The Eltonian food chain was a compelling idea and, even many years after it was first described by Elton, it was to inspire the clever title of a book of essays by the ecologist Paul Colinvaux, *Why Big Fierce Animals Are Rare* (1978).[14] All the lions, leopards, and cheetahs, indeed all of the vertebrate predators of the African savanna, are vastly outnumbered and outweighed by the numerous grazing animals that collectively form their prey. Sparrows, which eat mostly grain, are generally numerous and small, while hawks, which eat mammals and birds, are far less numerous and larger, at least in comparison with sparrows.

How much energy moves from one component of a food chain to another? The task of solving this difficult problem fell to a young ecologist named Raymond Lindeman. In 1942, Lindeman published one of the most important papers ever to appear in the journal *Ecology*, with the title "The Trophic-Dynamic Aspect of Ecology."[15]

Lindeman attempted to simplify the complexities that seemed inherent in descriptive models such as those of Elton. He studied Lake Mendota (in Wisconsin) and Cedar Bog Lake (in Minnesota). He combined all of the photosynthesizing organisms, the plants, into one category (such categories are now called "trophic levels") that he called "producers." The organisms that all ultimately depend on plants as an energy base, the protozoa, worms, clams, and fish, he called "consumers." And the collective organisms, mostly bacteria and fungi, that consume waste, dead tissue, and recycled inorganic material back to the producers were termed "decomposers." As happens in science, Lindeman's paper was so revolutionary at the time that many ecologists initially failed to understand its importance. (The same result occurred in July 1858 when papers by Alfred Russel Wallace and Charles Darwin were first read at the Linnaean Society in London.)

Lindeman estimated the flow of energy through each of the trophic compartments of the ecosystems. He then compared the differences between trophic levels, calculating what he termed progressive efficiencies. In other words, what percent of the energy in a given trophic level actually passed into the next trophic level above it? For example, in Cedar Bog Lake, the progressive efficiency from producer to primary consumer was 13.3%. It was 8.7% in Lake Mendota. Averaging between the two lakes, about 11% of the energy passing through the producer trophic level actually enters the primary consumer or herbivore trophic level. This figure may seem low, but plants must devote much energy to growth as well as metabolism, energy of respiration, which is converted to heat and therefore unavailable to consumers. Beyond that, it is possible, of course, that predation from the higher trophic levels limits the overall numbers of herbivores (what ecologists now call a top-down effect), thus preventing additional depletion of the producers.

Lindeman provided the first quantification about how energy moves from one trophic level to the next and also demonstrated how the first and second laws of thermodynamics influence the trophic structure of ecosystems. That was a big step forward. Lindeman's work stimulated the development of a sophisticated research program based on models of ecosystem function. That's arguably not a paradigm shift, but it is nonetheless an essential insight into ecosystem structure and function. Lindeman thought "outside the box," and as such, eventually inspired other innovative research programs, particularly those of two brilliant brothers, Eugene and Howard Odum.

The Odum brothers focused their careers on studying how energy moves through ecosystems, and did it so well that they defined the direction of ecological research for many ecologists. Gene Odum's work incorporated the use of radioactive tracer elements to measure rates of energy and mineral flux in the field. Howard Odum's work was equally if not more broad in scope, measuring energy flow through whole ecosystems.

In perhaps one of the most ambitious ecosystem-level studies of the twentieth century, Howard Odum published a complete trophic analysis of the freshwater ecosystem of Silver Springs, Florida (which also was a nice place to work).[16] Comparing his results with those of Lindeman, Odum noted a 5% efficiency of transfer from Sun to producers. This was much higher than that measured by Lindeman, whose data showed efficiencies from Sun to producers to be a mere 0.1% (Cedar Bog Lake) and 0.4% (Lake Mendota). However, those two lakes were located far to the north and thus had a much shorter growing season. The Lindeman efficiencies were based on measures of gram-calories per square centimeter per year. The Sun shines a lot more in Florida over the course of a year, and daily temperatures average much higher than in Wisconsin or Minnesota.

Odum found a 16% transfer efficiency from plants to primary consumers (herbivores), an 11% efficiency from primary to secondary consumers (carnivores), and a 6% efficiency from secondary consumers to top carnivores. These efficiencies, which were each generally similar to those obtained by Lindeman, have come to be called trophic transfer efficiencies and have given rise to what ecologists term the "ten percent rule." More a guideline than a rule (and based on laboratory as well as field studies), it states that, on average, only about 10% of the energy present in a trophic level will pass to the next highest trophic level. This means that, on average, an ecological pyramid of energy decreases at each level by an order of magnitude.

Given this reality, due mostly to constraints imposed by the second law of thermodynamics,[17] it is clear why Eltonian food chains are normally limited to only about four to six trophic levels. There is simply insufficient energy to maintain any additional trophic levels. Even a tyrannosaur, at about forty feet long and weighing up to seven tons, was only three steps from the Sun. And if ever an animal was at the apex of a food chain, it was *Tyrannosaurus rex*.

Ecologists also recognize that other factors contribute to food chain length, including overall energy availability, dispersal and migration, and disturbance frequency. But lording over all other factors are the realities imposed by the laws of thermodynamics.

Ecosystem studies blossomed in the 1950s and 1960s, in part as a result of the availability of new methodologies pioneered by the Odums and others. Ecosystem ecology ruled.[18]

In the 1960s and 1970s, an ambitious project named the International Biological Program (IBP) involved both North American and European ecologists in attempts to utilize increasingly powerful computers to develop accurate mathematical models of whole ecosystems. IBP was successful in focusing the importance of research at the ecosystem level, although it was far less successful in achieving its high-reaching goals of developing reasonably accurate, predictive models of complex ecosystems. It was an admirable attempt that brought many talented graduate students into the field of ecology.

Not all ecologists focused their research on ecosystems. Many studied species interactions, investigating population biology and such things as intra- and interspecific competition. These studies, some of which were lab-based, some field-based, were largely driven by a debate about what factors regulated natural populations. At the center of the debate was, yet again, the balance of nature. Was there or wasn't there a balance of nature? On page 68 of the *Origin*, Darwin made an observation that was representative of what was at issue in the population regulation debate.

Darwin noted that the birds inhabiting the grounds around his home south of London experienced very high mortality during the winter months of 1854–55. He wrote, "Climate plays an important part in determining the average numbers of a species, and periodical seasons of extreme cold or drought, I believe to be the most effective of all checks. I estimated that the winter of 1854–55 destroyed four-fifths of the birds in my own grounds; and this is a tremendous destruction, when we remember that ten percent is an extraordinarily severe mortality from extraordinarily severe mortality from epidemics with man."

Darwin was describing what ecologists termed "density independent effects" on populations. In other words, it did not matter just how many birds were present around Darwin's property. What mattered was simply that the weather turned severe, likely too cold for most of the birds, no matter how many or which species, to endure. Darwin presumed they perished, though it is possible that some may have moved elsewhere in search of a more hospitable climate.

Weather is a major factor affecting populations. It is density independent because the weather event, whatever it is, is not sensitive to or determined in any way by the size of the population, whatever it may be. It is a totally extrinsic effect. Because they are extrinsic, density-independent factors are not considered to "control" growth of populations in any precise sense. In other words, there is no feedback loop between weather (or any other density-independent factor such as fire) and population growth. If, for example, a population of wildebeest in Africa were growing beyond its local carrying capacity, and if that fact caused a drought to ensue, forcing the wildebeest to cease reproducing, it could be concluded that weather regulates wildebeest reproduction. But such a notion is, of course, ridiculous, though weather certainly affects wildebeest populations as it does other populations on the Serengeti. Drought, an annual occurrence on the Serengeti, is responsible for the impressive animal migrations that characterize that region. Many animals perish during the dry season and thus climate affects populations, but in no precise manner.

Many ecologists, especially entomologists, were persuaded that density-independent effects were of monumental importance in affecting population density. But not all ecologists were so persuaded.

Laboratory populations of animals such as fruit flies (*Drosophila*), flour beetles (*Tribolium*), and various protozoa such as *Paramecium* were shown to grow such that they typically reached an equilibrium, a density at which birth rate roughly balanced death rate. As they approached this "carrying capacity," their rates of reproduction declined. Thus the rate of reproduction was a function of density of the population, called "density dependence." Is density dependence also typical of nature outside of the laboratory?

Ecologists engaged in a vigorous debate as to whether or not most populations in nature are fundamentally controlled by density-independent or density-dependent factors. Two seminal works focused the problem: *The Distribution and Abundance of Animals* by H. G. Andrewartha and L. C. Birch, and *The Natural Regulation of Animal Numbers* by David Lack.[19] Andrewartha and Birch, both entomologists, argued strongly for the overriding importance of density-independent factors affecting most populations, while Lack, an ornithologist, assembled evidence to support the importance of density dependence, particularly as it related to food limitation.

For example, Andrewartha and Birch discussed the life cycle of *Thrips imaginis*, a small insect that is considered a pest species on rose flowers. Though they fully agreed that the populations of this insect are limited by availability of rose flowers, their only food source, Andrewartha and Birch showed that the flowering of roses is highly affected by weather, thus weather ultimately regulates thrips populations. Lack, working with birds, showed that the clutch sizes of various species of birds vary with population density and food supply in a manner consistent with density-dependent population regulation.

I could spend the remainder of this book documenting examples of both density-independent and density-dependent population

studies of insects, vertebrates, and plants. Both density indepen-
dence and density dependence abound in nature and have been
well studied. They are not as distinct as might appear to be the
case. For example, in the case of a large population affected by a
severe weather event, the survivors might be only those that found
shelter. The density of the population thus becomes a factor in
overall mortality level.

The debate about density factors and their influence was fruitful
in that it led to insightful research. For example, David Lack, in his
elegant study of Darwin's finches of the Galápagos, showed that
interspecific competition among similar species leads to specializa-
tion and segregation of niche space, a generally density-dependent
effect and one with evolutionary consequences.[20]

G. Evelyn Hutchinson defined the ecological niche as an *n-
dimensional hypervolume*, a polysyllabic term that translates to
the notion that each species, indeed each organism, lives strictly
within its abiotic and biotic parameters (largely genetically and
thus evolutionarily determined).[21] The more similar those parameters are among species, the more likely there will be competition.
There is what has become known as a *fundamental niche* for each
species, the physiological limits within which the species exists.
But to the extent that different species require the same resources
(abiotic and biotic), they compete, leading to the notion of *realized
niche*, the "ecological space" actually occupied by a species. From
Hutchinson's concept of the ecological niche came fruitful theoriz-
ing on just how similar animals could be and still coexist, without
competition eliminating some species.

In one study, which is considered a classic, five species of wood-
warblers (Parulidae) were studied in boreal forest.[22] The researcher
was Robert MacArthur (1930–72), who proved to be such a force
in ecology that he influenced research programs long after his pre-
mature death.

MacArthur showed that each of the five bird species foraged
most of the time in a specific part of the tree or in a manner slightly
different from each of the other species.[23] While the five species
appeared to be foraging together in the same conifer, in truth they
were segregating. The genius of MacArthur's study was that he

took a detailed, statistical look at what seemed to be a situation in which five species had strongly overlapping *foraging niches*.[24] Anyone looking at a white spruce stand can, with patience, record each of the five warbler species foraging in the tree. By timing exactly where and for how long each bird foraged, MacArthur was able to show that the Cape May warbler (*Dendroica tigrina*) usually forages near the tops and the black-throated green warbler (*D. virens*) forages mostly along the outside middle part of the tree. The bay-breasted warbler (*D. castanea*) also forages in the middle part of the tree but more toward the inside and center than the black-throated green. MacArthur's data represent statistical truths. One can, indeed, see a Cape May warbler in the middle of a spruce tree, but that happens only about 5% of the time. About 58% of the time, the species is found foraging at the treetop.

What MacArthur's study demonstrated became known as niche segregation. The foraging niche of the five wood-warbler species was subdivided such that each of the five had "exclusivity" along some part of the niche axis. It was reasonable to assume, as MacArthur did, that interspecific competition had acted to narrow the foraging niche of the birds, resulting in their coexistence rather than competitive exclusion. Note that this is a strongly evolutionary explanation, as it implies that competition among the bird species, once intense, is now less intense, as each has specialized to its own unique foraging niche. Evolutionary explanations had come fully into ecological analysis. The ecological embrace of evolution was fully realized in work on life history studies beginning in the 1960s. Demographic theory based on evolutionary processes had its origins in studies such as MacArthur's and is a burgeoning research area in ecology today.

With the evolutionary breath of fresh air breathed into the field by Lack, MacArthur, and others, ecologists moved ahead quickly. For example, ever since Darwin it was realized that many species "coevolve," which means that each influences the evolution of the other, sometimes resulting in a total interdependence, a mutualism, between the two species (such as when specific plant species depend on specific insect pollinators). Ecologists soon learned that evolutionary history and the process of natural selection all profoundly affect mutualisms, as well as relationships between predators and prey, between parasites and hosts, and between pathogens and hosts. The ecology of disease and pathogens is currently an area of intense ecological study.

Among the most seminal studies was one also spearheaded by Robert MacArthur, along with his colleague Edward O. Wilson. In 1967 they published a monograph, *The Theory of Island Biogeography*, a work that combined mathematical modeling with both ecological and evolutionary data.[25] The MacArthur-Wilson model demonstrated that islands (and anything from an isolated woodlot to a continent can be considered an "island" of a sort) reach a theoretical dynamic equilibrium in species richness, where rate of immigration (or speciation) equals rate of emigration (or extinction). The equilibrium richness is based on the area of the island and its distance from colonization sources. It is essential that the concept of equilibrium not be viewed as the same as a balance of nature. It is highly dynamic, subject to frequent change. To say it is a "balance" has little real meaning. MacArthur and Wilson recognized this clearly. The value of their study, which has been tested in numerous contexts with variable results, allows researchers to measure the species richness (of birds, or plants, or whatever) in the relative context of its potential theoretical richness.

With the availability of digital computers beginning in the 1960s, mathematical models were increasingly used to study species interactions. Initially virtually all models called for some sort of equilibrium to develop (as indeed, an assumption of a balance of nature would dictate); but later models, and those in use today, added complexity, included stochastic factors, and did not assume an equilibrium as an end point. Thus the balance of nature as a concept was eroded as these nonequilibrium models explained data best.

The field of ecology burgeoned from the 1970s on, following the first Earth Day (April 22, 1970), when ecology more or less officially joined the vernacular of "household words" (chapter 14). Ecology research, once largely confined to ecosystems of the temperate zone, expanded to encompass both polar and tropical areas as well. The increased ease of travel greatly enhanced the ability of ecologists to visit field sites that, just decades ago, would have

been difficult to access. With travel has come a greater apprecia-
tion of the world's biodiversity, and thus ecologists are on the fore-
front of biodiversity conservation (chapter 13).

Today many ecologists are investing their energies in Long Term
Ecological Research (LTER) studies that utilize a common database over many years.[26] Only through such studies can large spatial and temporal scale effects be detected, thoroughly documented,
and ultimately understood.

Scale in space and time, as you will soon learn, now provides an
overarching framework for ecology (chapter 7). No professional
ecologist that I know would now say there is a balance of nature.[27]
But it required over a century from the time of the publication of
the *Origin* for that baggage to be discarded.

A Visit to Bodie
Ecological Space and Time

Ancient Greek philosophy can be found in unusual places, including on a wall in a ghost town. Situated among the hills just east of the extensive Sierra Nevada mountain range in the Great Basin Desert of California, the town of Bodie was founded in 1859, the year Charles Darwin published *On the Origin of Species*. In that year a certain William "Watermelon" Bodie succeeded in his search for gold, at a place that became known as Bodie Bluff. By 1880 some 10,000 people inhabited the thriving town of Bodie, which rose from scratch among the sagebrush and hills of desert. Bodie was anything but genteel, inhabited by gunfighters and prostitutes who frequented opium dens, gambling houses, brothels, and the town's reputed 65 saloons.[1]

Bodie thrived for some years, not many, but some. Its resource base, to use an ecological term, soon was exhausted. It also suffered from unpredictable effects of nature. An avalanche destroyed the town's power plant in 1911, and the second of two major fires destroyed much of the town's buildings on June 23, 1923. The town never recovered. Some buildings remained after its abandonment, and in 1964 the town that once hosted hundreds of very rough and tough humans became a California historic park. Tourists now walk the old streets of the town. There are no opium dens, but a few places do sell postcards.

You have to want to find Bodie, as its location is not exactly on the tourist trail. It is at the termination of a thirteen-mile secondary road that winds through the desert from California Highway 395, well north of Mono Lake and the entrance to Yosemite National Park. I wanted to see it and so did my wife Martha and stepdaughter Sarah, so we went and we toured.

In its heyday Bodie could not lay claim to being an "intellectual" town. Nonetheless, in one of the decrepit buildings, I came upon a sign that read "Nothing Endures But Change." It identified the quote as from Heraclitus (c. 540–480 BC), a Greek philosopher. The sign kind of summed up the history of Bodie—as well as providing an apt description of ecology.

As I left that old building I came upon a greater sage-grouse (*Centrocercus urophasianus*), resembling a large chicken, as it was poking around in a clump of sagebrush. I thought about the sage-grouse generations that had lived through Bodie's short history and all that had lived before and since. Bodie came and went quickly. The desert hills came earlier, after the rise of the Sierra Nevada mountain range, and will eventually disappear, but that will take longer. The greater sage-grouse will also eventually become extinct, a ghost bird that once inhabited a ghost town.

There is a seeming constancy about any animal or plant species. Greater sage-grouse will look, sound, and act like greater sage-grouse throughout any human lifetime. John James Audubon, in his western perambulations, observed sage-grouse that were not phenotypically distinct from those alive today.[2] Such constancy initially made creationism seem logical and organic evolution implausible. The Darwinian Revolution reversed the focus from species stasis to mutability, with the resultant evolution of new forms from previously existing forms. One of Darwin's major arguments in attempting to persuade his readers of the truth of evolution was that the process is too slow to be easily detected within human lifetimes. The normal time scale of evolution, like that of geology, generally exceeds the units of time that are intuitively meaningful to humans: minutes, hours, days, years, decades. Thus organic evolution is far from obvious in everyday human experience. It is not unlike watching an hour hand on a clock: it does not appear to move when watched constantly but only "moves" when examined at appropriate intervals.

Ecosystems are perceptually like species in that they too exhibit characteristic form (physiognomy), impressive complexity, and apparent constancy. These perceptions are captured in the notion that nature is somehow balanced. An oak-hickory forest, a northern hardwood forest, and a spruce-fir forest each have distinct and easily recognizable characteristics that make it possible for virtually anyone to distinguish among them.[3] A cactus-dominated desert such as the Sonoran Desert is distinct from a shrub desert such as the Great Basin Desert. These ecosystems, unless radically disturbed, will persist well beyond any human lifetime and thus appear to be stable. From apparent stability, with a good historical dose of Greek philosophy, comes the belief that such ecosystems are also intrinsically balanced.

In the naïve sense, "balance of nature" implies total interdependency, the notion that species are like dominoes arranged in such a manner that if one falls, others, maybe many others, inevitably follow. Another analogy is that of a machine. Start taking away or adding parts and soon the machine malfunctions. But nature is not a machine, and we should beware of relying too much on reasoning by analogy. To ecologists, the notion of balance of nature is more apt to be expressed as the belief that ecosystems eventually attain equilibrium, becoming self-reproducing and resistant to invasion by other species (chapter 13). In this view niche space is allocated among constituent species, such that competition among species is minimized or nonexistent. The equilibrium concept is akin to Darwin's analogy of the wedge, each species "wedging" itself into a tightly structured ecosystem. Further, a balance of nature can be envisioned as a kind of normal "set point," a species composition to which disturbed ecosystems return in the process of "recovery" following perturbation.

Forests, deserts, and prairie, as well as many other ecosystems, have been cited in various writings as examples of the balance of nature. By way of example, let me focus on the vast eastern deciduous forest, the ecosystem in which I did my graduate work, one of the kinds of ecosystems I know best. It dominates eastern North America, from the Mississippi River to the Atlantic Coast, from the Gulf States into Canada. But mostly, I will focus on New England.

Henry David Thoreau saw order in the way in which abandoned New England pastures would gradually revert back to forests.[4] Years later, Frederic Clements (1928), whom we met in the previous chapter, would discuss this same process, ecological succession, as an example of how ecosystems can be viewed as "supraorganisms," in which vegetation development (also called ecological succession) is sufficiently orderly to be analogous to life cycle changes in organisms. According to this view, vegetation development terminates in a mature, stable (balanced) ecosystem typical of the climatic region. Later still, Eugene Odum (1969), in one of the most influential papers in modern ecology, would envisage ecological succession as a natural "strategy" of nature to restore complexity and stability (and implied balance) to ecosystems.[5]

Well-meaning conservationists continue today to argue against ill-conceived assaults on ecosystems as interfering with the balance of nature. Arguments are occasionally heard in which it is said that "nature knows best," and should be left to its own devices. Such a view is usually based upon the presumption of a natural balance. But if there is no natural balance, such arguments need to be reformulated.

The concept that there is, indeed, a natural balance of nature has been viewed somewhat differently throughout history, but is nonetheless very deeply ingrained in human thinking.[6] And it persists.

One of the more controversial hypotheses about nature's presumed intrinsic balance has to do with biogeochemical cycling. As the name "biogeochemical" implies, elements such as calcium, phosphorus, nitrogen, and carbon all routinely pass between the living or biotic components of ecosystems and the nonliving or abiotic components. The precision with which these elements are cycled and recycled led James E. Lovelock of England to propose a unique idea.[7] He called it Gaia, a name that, almost immediately, stimulated intense interest, enthusiastic support, outspoken criticism, and some ridicule. Lovelock adopted the name Gaia from the ancient Greek mythological goddess of the Earth (always a dangerous practice in my opinion, to humanize what is intended to be a scientific proposal). His use of the name was based on his contention that Earth, taken on the broadest scale, can be metaphorically visualized as an immense living organism unto itself that has evolved elegant and largely microbial mediated feedback loops that ensure that it remains homeostatic. In other words, Earth regulates itself to maintain the atmosphere at its present concentration of 21% oxygen, 79% nitrogen, as well as the constancy of such variables as the salinity of the oceans, the temperature range of the planet, and so on. Lovelock offered some models that purportedly demonstrated how systems lacking any sense of planning or intelligence could, nonetheless, behave in such a manner as to produce negative feedback loops that would maintain such variables as oxygen concentration and temperature within relatively narrow limits (and thus prevent "runaway" processes such as unrestrained carbon dioxide buildup, as occurred on the planet Venus). The principal actors in Lovelock's world of Gaia are microbes, whose cumulative global effects are most responsible for keeping biogeochemical cycles homeostatic. No ecologist would dispute this reality.

Criticism of Lovelock's notion of Gaia was immediate. One weak criticism centers around the name itself, Gaia. To some critics it suggests that Earth is a huge "organism" with its own innate form of "intelligence," a conclusion repugnant to most scientists. This view was apparently never Lovelock's intention. More reasoned criticism centers around whether or not evolutionary principles could and would result in the development of elaborate feedback loops demonstrable on a global basis. Ecologists generally hold to the view that, while populations of organisms evolve, ecosystems do not evolve, at least not in any actual Darwinian sense. Ecosystems change because organisms evolve as Earth's climate changes. Carbon dioxide and other greenhouse gases are now strongly implicated in the present alteration in global climate, producing an overall trend toward warming. History shows that climate change has been typical (chapter 11), though, as Lovelock suggests, the composition of the atmosphere has remained within relatively narrow limits.

In the present century, if humans succeed in reversing or attenuating the effects of increased greenhouse gases, it could be argued that Lovelock is fundamentally correct, and that humans have assumed the role that fell mostly to microbes throughout the millennia, that of stabilizing the planet's essential life supporting systems. I guess we'll have to wait and see.

Now use your imagination and consider those vast eastern forests that must have met the eyes of the Pilgrims in 1620. That scale of time is within recent human history. But think back also to before humans even occupied North America, indeed, before humans even existed. How permanent is a forest?

First, at the risk of stating the obvious, there was no eastern forest for most of the 4.5 billion-year history of the planet. The evolution of flowering plants and of the various species of birds and mammals that live among them, of oaks and pines, of jays and wood-warblers, at most encompasses less than two percent of the total history of Earth. Any hectare of land that now supports an oak-hickory forest once was barren and lifeless. It required many millions of years before the land became populated by tree-sized horsetails and primitive conifers, among which lived early dinosaurs. Only in the last years of the Mesozoic era and throughout the Cenozoic era, a time span of at most 80 million years, have modern conifers and broad-leaved, flowering trees and shrubs been present (though certainly not in their current species assemblages).

And the means by which multitudes of evolutionary changes have been effected is far from orderly. The Mesozoic era is now widely believed to have ended abruptly following an impact by a large asteroid (chapter 10). The Cenozoic era has seen Earth's climate becoming increasingly temperate and less equitable, with relatively recent periods of major glacial advance. During the height of recent glaciation (about 25,000 years ago), up to two miles of ice sat atop land that now supports coastal oak-pine forest. Forests extant prior to the glacial advance were fragmented, their component species displaced southward to various ice-free "refuges." When the ice retreated, the forest did not migrate back as a unit, but rather individual species moved northward at different rates.[8] Such factors as dispersal mechanisms, seed size, and, most importantly, climatic characteristics each influenced the rate at which each plant species expanded northward following glacial retreat (reflecting a very Gleasonian view of the plant community).

The present composition of eastern forests is truly "modern" in origin, dating back only to the few thousand years of the present interglacial period, and continuing to change up to and since the time of European settlement. It is also largely coincidental, with species assemblages representing various combinations of those species with sufficient dispersal powers, physiological hardiness, and adequate competitive abilities to persist together. To reiterate, this is a view of the ecological plant community originally put forth by Henry Gleason and referred to as the "individualistic plant community."

Animal communities, like plant communities, demonstrate individualistic assembly at various temporal and spatial scales. Beavers (*Castor canadensis*), porcupines (*Erethizon dorsatum*), and bobcats (*Lynx rufus*) all presently cohabit large tracts of northern hardwood and boreal forest. But beavers first evolved in Europe about 35 million years ago, arriving in North America only about 15–20 million years ago.[9] Some beaver species, now extinct, did not even inhabit forests. Ancestors of the porcupine, presently a common rodent throughout much of northeastern forests, originated in South America, moving northward along with species such as sloths, armadillos, and opossums during the great faunal interchange that occurred between North and South America when glaciation exposed the Isthmus of Panama.[10] Felids, like beavers, appear to have evolved in Europe, about 24 million years ago, and first arrived in North America around 18 million years ago. Examined on a long-term time scale, the various species that today form natural associations are really accidents of history.

Such processes continue today. Currently the following largely unrelated bird species are expanding their ranges northward, and, in Massachusetts, each has undergone significant population increases in the last thirty years: turkey vulture (*Cathartes aura*), red-bellied woodpecker (*Melanerpes carolinus*), Acadian flycatcher (*Empidonax virescens*), tufted titmouse (*Parus bicolor*), Carolina wren (*Thyothorus ludovicianus*), blue-gray gnatcatcher (*Polioptila caerulea*), worm-eating warbler (*Helmitheros vermivorus*), northern cardinal (*Cardinalis cardinalis*), and orchard oriole (*Icterus spurius*).[11] A few mammalian species, such as Virginia opossum (*Didelphis virginiana*), are also undergoing significant range expansion northward. As implied above, these are all basically southern species undergoing range expansion. I will discuss this trend further in chapter 11. Just note here that all of the species mentioned above are quickly becoming common where, decades ago, they were rare. Ecosystems are anything but closed to immigrant species. Darwin's "wedge analogy" does not apply.

The rapid population growth of such species as European starling (*Sturnus vulgaris*), house sparrow (*Passer domesticus*), and house finch (*Carpodacus mexicanus*), none of which is native to the Northeast, also supports the notion that ecosystems are not so naturally balanced as to close out opportunities for invader species. Indeed, the ease with which some species colonize is remarkable.

The contrary is also true. The twentieth-century loss of the American chestnut (*Castanea dentata*) as a numerically dominant tree species throughout eastern North America resulted in a negligible ecological affect on the forest as a whole. Oaks and hickories fundamentally replaced it as numerically dominant tree species throughout the chestnut's range. The most numerous bird species ever to occur in northeastern forests was the passenger pigeon (*Ectopistes migratorius*). The species was almost certainly an important disperser of chestnuts, oaks, and hickories, but it is now extinct and, importantly, there is no evidence that its loss has resulted in any significant reduction in mast crop producers throughout its former range.

Smaller temporal and spatial scales are also illustrative of the fact that nature is dynamic, not static.

As a basic tenet of how nature is, consider this: If a habitat, any habitat, is left entirely alone, protected, with nothing done to it, it will nonetheless eventually exhibit change. Change is inevitable because eventually some form of natural disturbance will occur, climate may alter, new species will invade, extant species will drop out.

Natural disturbance is now recognized by ecologists as the primary factor that maintains ecological communities in a nonequilibrium state. Because disturbances, both temporally and spatially, tend to overlap, the overall view of community change has been termed *hierarchical patch dynamics*.[12] Think of a patchwork quilt where the various patches are of differing ages and sizes, where they change size and come and go in a kaleidoscopic pattern. The quilt is never uniform and never entirely static. Change is always happening somewhere. What varies is the area changing and the rate of change. These variables are considered together as scale effects. That is how ecologists now view nature. There is no room for balance of nature is such a view.

Even such complex and species-rich ecosystems as lowland tropical rainforests owe much of their vast biodiversity to multiscale, periodic disturbance.[13] Disturbance may range from modest to severe. What ecologists have learned is that intermediate disturbance levels seem to result in maximum biodiversity. Too little disturbance and competition among species will eliminate some species and reduce the diversity. Too much disturbance and few species will be able to tolerate the frequency of perturbations.

Regarding area, a disturbance may be small scale, such as a single treefall caused by a lightning strike; moderate scale, such as a localized blowdown of a group of trees; or large scale, such as a widespread fire or the effects of a major hurricane. American beavers are historically responsible for much forest disturbance at intermediate levels. Their habit of cutting trees and damming streams ultimately creates expansive open meadows where many species thrive that could not inhabit closed forest.

In the Northeast, the presence of successional species such as most goldenrods (*Solidago spp.*), asters (*Aster spp.*), sumac (*Rhus spp.*), eastern red cedar (*Juniperus virginiana*), field sparrows (*Spizella pusilla*), prairie warblers (*Dendroica discolor*), and brown thrashers (*Toxostoma rufum*) is the evolutionary and ecological result of relatively frequent disturbances providing a continued presence of open, nonforested areas.[14] These species, though undoubtedly less abundant when the region was densely forested, are nonetheless native to the region, just as closed-forest species such as sugar maples (*Acer saccharum*), eastern hemlocks (*Tsuga canadensis*), and wood thrushes (*Hylocichla mustelina*) are.

The Pleistocene arrival of *Homo sapiens* was of profound significance in the effect it had on North American ecosystems. Because culture provides humans with the ability to greatly alter nature, and because such alterations often result in extreme change to ecosystems, it is understandable that humans view themselves as having disturbed nature's natural balance. Particularly in Western culture, humans perceive themselves as largely apart from nature,[15] a dualism that isolates humans from nature as well as often putting them at odds with nature (chapter 14). Even Native Americans, a group of people often collectively identified as having a deep-seated cultural kinship with nature, nonetheless exerted significant and often negative effects on natural ecosystems.[16] Indeed, evidence exists that the immigration of people into North America soon produced a devastating effect on the so-called Pleistocene megafauna such as the ground sloths and mammoths, resulting in a massive wave of extinctions.[17]

In general, it is believed that the overall abundance of eastern forest wildlife dramatically declined as populations of European settlers took increasing control over the landscape. Various early estimates of animal abundance suggest vast concentrations of wildlife.[18] However, it is difficult to make accurate comparisons of largely anecdotal data taken by a select few observers with such modern coordinated databases as the Breeding Bird Survey.[19] For example, both John James Audubon and Thomas Nuttall believed the bay-breasted warbler (*Dendroica castanea*) to be extremely rare, and Audubon doubted that it bred in the United States, writing that it "must spend the summer in some of the most remote north-western districts, so that I have not been able to discover its principal abode."[20] Audubon commented on the extreme rarity of the chestnut-sided warbler (*Dendroica cantanea*), a species he encountered but once. Both species are very common today. Given that both Audubon and Nuttall worked without binoculars, it is certainly possible that they may have somehow overlooked these species. But given their obvious observational skills with numerous other species, it seems unlikely. More likely is that the bay-breasted and chestnut-sided warblers were really rare and both have greatly increased in population only during the present century.

No one doubts that humans have been responsible for several large-scale perturbations affecting eastern forest bird and mammal communities. The result is that forests today are markedly different from those present at the time of European colonization. The most significant of the many ecosystem alterations occurred in the eighteenth and nineteenth centuries as European settlers cleared forest for pasture, agriculture, and human habitation.[21] Forest-dwelling species were faced with a significant decline in available habitat. The one bird species that probably suffered the most from forest clearance might well have been the passenger pigeon, once estimated as perhaps the most abundant bird species on Earth. The extinction of the passenger pigeon, often assumed to have resulted from excessive hunting pressure, was probably more the result of extreme habitat loss from forest clearance.[22]

As an obvious result of forest clearance, grassland and successional species were able to expand their ranges throughout the eastern states as forest species presumably became collectively less abundant. Species such as eastern meadowlark (*Sturnella magna*) and grasshopper sparrow (*Ammodramus savannarum*) undoubtedly benefited from forest clearance, just as populations of these species have recently suffered from habitat loss partly caused by regeneration of forest throughout the twentieth century. Indeed, ecological succession of field to young woodlot is considered a significant factor in the recent decline of the golden-winged warbler (*Vermivora chrysoptera*), which inhabits early successional habitats, and its replacement by the similar blue-winged warbler (*V. pinus*), which can inhabit a wider range of disturbed areas, including late successional deciduous forests.[23]

Following European settlement and the growth of European populations, predator species such as mountain lion (*Felis concolor*) and gray wolf (*Canis lupus*) were rapidly extirpated from eastern regions. In 1717 gray wolves remained sufficiently common on Cape Cod, Massachusetts, that it was proposed to build a fence between Sandwich and Wareham to exclude wolves and make the outer Cape a livestock sanctuary.[24] The fisher (*Martes pennanti*), a species now very much on the increase in southern New England, was greatly reduced by fur trapping. In general, the collective effect of such activities was to significantly deplete top carnivores from eastern forest ecosystems, a trend that persists today, though some species such as red (*Vulpes fulva*) and gray (*Urocyon cinereoargenteus*) foxes, raccoons (*Procyon lotor*), and coyotes (*Canis latrans*) appear to be thriving, and other species, such as the fisher, are rebounding at least in parts of their ranges. White-tailed deer (*Odocoileus virginianus*) populations, often problematically abundant today, owe much of their present abundance to the absence of large predators.

Now consider what you just read. Deer have increased in the absence of natural predators. So does that represent an imbalance of nature? Most would readily say yes. But what then is the optimum density of white-tailed deer? How many predators should there be? Doesn't much of the answer ultimately depend on the quality of plant food available to the deer? Won't deer populations that become dense be more susceptible to bacterial and viral pathogens, or perhaps parasites, and won't that "naturally" lower the population density? Why are deer populations "balanced" by hunting pressure but not from eventual starvation due to loss of plant food? Do predators such as wolves and coyotes balance the deer population and pathogens such as bacteria and fungi not? I will discuss these sorts of questions in chapter 12.

Human actions continue to alter the animal communities of eastern forests. Avian predators such as the accipiter hawks (*Accipiter spp.*), peregrine falcon (*Falco peregrinus*), and soaring hawks (*Buteo spp.*), have increased with protection and conservation awareness and, in the case of the peregrine, captive propagation. Wild turkey (*Meleagris gallopavo*) populations have been successfully reestablished throughout much of their former eastern range. But at the same time, habitat fragmentation largely due to increasing human populations in suburbia poses threats for some species such as wood thrushes, while others, such as blue jays (*Cyanocitta cristata*) and common grackles (*Quiscalus quiscula*), seemingly benefit. Things just keep changing.

Eastern forest ecosystems continue to change; some species decline, others advance. As a result of research over the past several decades, ecologists have come to understand the reality of ecosystem dynamics, and have largely abandoned the notion that nature exists in some sort of meaningful natural balance. What effect should this reality have on informing decisions regarding land use, conservation of species, and overall stewardship of planet Earth? I will let that question incubate in your mind and will take it up again in the final chapter when we meet Marley's ghost.

8

Ecology and Evolution
Process and Paradigm

G. Evelyn Hutchinson authored a classic book about ecology with the title *The Ecological Theater and the Evolutionary Play*.[1] Hutchinson was a genuine polymath when it came to ecology, and a good few of his students became some of the most eminent ecologists of the twentieth century.[2] Hutchinson's brief book, written (as this book is) for the lay reader, is engaging as it describes evolution as an ongoing process shaping Earth's various ecosystems, whatever they might be. But there is a deeper, more profound meaning. Evolution is the paradigm, not ecology. Ecology is the process, and it provides the actors and the setting, the "stage." Ecology continues to enjoy many an empirical success in describing details of how ecosystems are structured and function, but these triumphs are essentially victories of empiricism, not paradigm revelation. All of ecology is really absorbed by evolutionary biology, particularly by natural selection theory. It is the purpose of this chapter to make that crystal clear.

I am part of a generation of ecologists who were introduced to the formal study of ecology by Eugene P. Odum's text, *Fundamentals of Ecology*.[3] This text was originally published in 1953 (384 pages) and I used the second edition, published in 1959 (546 pages). This book was energizing in more ways than one. First, it energized the reader, exciting me about how science is used to study nature. Second, it seemed to be all about energy: how energy moves through ecosystems, how energy structures food chains, how energy flows while materials cycle, how energy in the form of biomass accumulates with ecological succession. What the book lacked was any grounding of ecology in evolution. Evolution was relegated to just two topics, its influence on mutualism (when two or more species evolve to be interdependent for their mutual welfares), and paleoecology (how the recent fossil record can be used to reconstruct past environments). Darwin was mentioned twice, and natural selection was only briefly discussed in Odum's treatment of mutualism and how "cooperation" could evolve in nature.

The fifth edition of Odum's venerable text, coauthored with Gary W. Barrett, was published in 2005, fifty-two years after the first edition, and after Odum died in 2002 at the age of eighty-eight.[4] The treatment given evolution in the fifth edition is far more complete and sophisticated than was the case in the earlier editions. And that is as it should be.

Odum's 1953 text successfully eclipsed other early ecology texts, including a massive tome *Principles of Animal Ecology* (1949), coauthored by W. C. Allee, A. E. Emerson, O. Park, T. Park, and K. P. Schmidt. Each was an eminent ecologist, and their text ran a whopping 837 pages including index. The final section of the book was devoted to "Ecology and Evolution" and included discussion of genetic variation, adaptation, and natural selection. It also included a speculative section titled "Evolution of Interspecies Integration and the Ecosystem." Here the authors boldly hypothesized that Darwinian evolution might not be confined merely to organisms but to whole ecosystems. They wrote, "The evolution of interspecies integration involves the genetic modification of the ecologically associated organisms in relation to each other, in the aggregate resulting in the evolution of the community as a whole."[5]

Allee et al. thus viewed evolution as the process attaining an eventual balance of nature. Their conclusion makes this starkly apparent: "The probability of survival of individual living things, or of populations, increases with the degree with which they harmoniously adjust themselves to each other and their environment. This principle is basic to the concept of balance of nature, orders the subject matter of ecology and evolution, underlies organismic and developmental biology, and is the foundation for all sociology."[6]

Odum may not have done enough to include evolution, but Allee et al. did way too much. It was not until Robert Leo Smith challenged Odum's text in 1966 with *Ecology and Field Biology* that evolution and natural selection received appropriate treatment in a mainstream ecology text. Subsequent texts, and there are now many, all have a strong evolutionary focus.

Why should ecology be so dependent on evolution, particularly natural selection, for its explanatory power? When I first studied Odum's text I was quite satisfied that I was learning much even without any significant attention to evolution. But what I was missing, though I did not realize it at the time, was a crucial context, that of adaptation and natural selection. As I read Odum on ecological succession, for example, I simply accepted the description that annuals come first, then biennials and perennials, then grasses and shrubs, and so on. But why should such a pattern emerge? This chapter looks at some "whys." Consider, as a first example, a simple bottle of hot sauce.

A friend, knowing my taste for spicy food, gave me a bottle of Kali Kali, labeled as "Africa's hottest chili sauce." Cuisines all over the world feature chilies, chopped, whole, or in sauces. The hot sensation of taste that one experiences when eating chilies results from a chemical called capsaicin. As most of us know, a little of it goes a long way. The genus name for the chili plant, *Capsicum*, refers to this chemical. So why do some plants synthesize capsaicin?

Plants use energy for, among other things, making fruits, some of which are eventually consumed by various animal species, and some of those disperse the seeds within the fruits. Chili plants are no exception. Fruits contain the next generation of the plant, the plant's reproductive investment in the future of its genes. Various animals eat fruits but don't digest the seeds inside. When an animal later leaves droppings (or regurgitates), it disperses the undigested seeds, inadvertently performing a function essential for plants. Many plants have evolved mechanisms to attract seed dispersers. The ripening of fruit such that it is clearly visible on the plant is but one example. Seed dispersal is essential for Darwinian fitness!

Chili plants have evolved in such a way that the best seed dispersers are certain birds that don't seem to mind liberal doses of capsaicin. Mammals would also readily devour chili fruits and seeds but they are discouraged by the capsaicin. Thus birds eat the fruits and disperse chili seeds in their droppings, while mammals tend to avoid the fruits altogether. The capsaicin is an adaptation that keeps mammals at bay.

How do we know this, and, more to the point, why should it be? Two ecologists, Joshua Tewksbury and Gary Nabhan, have studied the relationship between chili peppers (the fruits of the chili plant) and the animals that eat them.[7] Their study was performed in southern Arizona and focused on the chili *Capsicum annuum* var. *glabriusculum*. The researchers asked how birds and mammals differ in their consumption of chili fruits and what happens to the chili seeds. Remote video observation of what kinds of animals fed on chilies during the day showed that one bird species, the curve-billed thrasher (*Toxostoma curvirostre*), accounted for 72% of all chili fruit removal. No mammals were seen on videotape eating chilies during the day. But then again, it is hot in the desert and many mammals are generally not out and about during the day.

Because most desert mammals are nocturnal, the researchers presented equal amounts of hot chili fruits and a bland fruit, desert hackberry (*Celtis pallida*), on sites available day and night to both birds and mammals. During the day, both kinds of fruits were removed, but at night, when mammals are active and birds are not, only hackberry fruits were taken. Mammals appear to want fruits but not when they are spicy.

In a follow-up experiment, researchers captured cactus mice (*Peromyscus eremicus*) and packrats (*Neotoma lepida*), as well as curve-billed thrashers. They offered the captive animals three food choices: the hot chili from the study site, a mutant variety of chili that is not hot, and hackberry fruits. The mammals readily consumed the hackberry fruits and took some of the mild chili fruits but refused to consume the hot chili. The thrashers ate all of the three kinds of fruits, including the hot chili. Furthermore, and importantly, of the seeds consumed by mammals, none of the mild chili seeds germinated, whereas the seeds of both hot and mild chilies consumed by the thrashers germinated as well as seeds carefully planted by the researchers. Passing through a bird's intestinal system does not harm the seed, but such is not the case with seeds ingested by mammals. The fact that mammals harm the seeds is a selection pressure on the plant population to evolve traits that favor birds as fruit consumers and thus seed dispersers.

The experiments, which are fairly typical of how ecologists do their research these days, showed that, in the case of chilies, common desert mammals such as various rodents are poor seed dispersers compared to birds. But this need not always be the case. In ecology, there are few sweeping generalities. Some mammals are effective seed dispersers. For example, bats are excellent seed dispersers for many kinds of seeds, particularly from tropical rainforest trees. Like birds, bats fly, and flight is the essential act that disperses the seeds. Many seeds pass through bat intestinal systems and still germinate. Thus many tropical plants have not evolved defenses that discourage bats. There is no selection pressure to do so.

In the Arizona study although less than 40% of the study area was shaded, 86% of adult chili plants grew in shade. This was due to the fact that thrashers typically deposit (via droppings) chili seeds in shade, where the birds rest during the heat of the desert day. Being dropped in shade is advantageous to the seeds as well, because they tend to suffer less from water loss when they eventually germinate. Tewksbury and Nabhan conclude, "In short, directed dispersal by thrashers to sites under bird-dispersed plants is highly beneficial to chilies." So, from the perspective of the chili plant, thrashers are the optimal seed dispersers.

Capsaicin is a chemical adaptation that confers evolutionary fitness on the chili plants in that it retards consumption of seeds by mammals, which are poor seed-dispersers, and permits dispersal by birds, which are good seed dispersers. Many questions obviously remain, such as how birds evolved tolerance for capsaicin in the first place (or even if they even needed to). Over the course of evolution, those chili plants that evolved capsaicin enjoyed a reproductive advantage over those that did not, and thus capsaicin became widespread and common in chilis. This is the process of natural selection, the very essence of Darwinism. An ecologist naïve about natural selection would have a dilemma to solve. Why shouldn't mammals take chili fruits? Why does capsaicin exist? Lots of "whys," all basically solved by context, the context of natural selection.

Why does the universe exist? Why is there a law of gravity? Why was I born? These "why-type" questions are grist for philosophers and metaphysicians. They have no place in empirical science because they are not knowable through scientific methodology. But, as shown above, a form of why-type question does, indeed, nest within science, and that has to do with natural selection and adaptation. As the evolutionist Ernst Mayr made clear,[8] the study of evolution involves two kinds of questions, how-type and why-type questions. How-type or "proximate" questions are the typical foci of laboratory science, questions answered with traditional scientific method. Why-type or "ultimate" questions are contextual, the context being adaptation by natural selection. Here's an example.

It is clear that some forms of plants such as sundew (*Drosera spp.*) and Venus flytrap (*Dionaea muscipula*) are "insectivorous." These plants have evolved unique ways to capture and digest insects. Plant physiologists understand how the plants function to accomplish what they do. But why are some plants insectivorous? That is not a metaphysical question but a real scientific inquiry, an evolution-based "why-type" question. Insectivorous plants generally grow in acidic bogs. The high acidity level retards bacteria from decomposing organic matter (as evidenced by ancient human remains preserved and "mummified" in northern European bogs). Because decomposition is compromised, nitrogen and other vital elements are in short supply, locked up in undecomposed peat. The insectivorous plants obtain nitrogen and other elements from the trapped insects, rather than by the more traditional route of the soil. In places where acidity levels are not so high as to prohibit bacterial decomposition of fallen organic matter, there are no selection pressures on plants that favor the evolution of insect capture and digestion. And there are no insectivorous plants. So to even begin to comprehend the ecology of insectivorous plants, one must understand evolutionary context.

Why-type questions also illuminate another reality of natural selection. Not all traits are, indeed, adaptive, at least not in any direct sense. Why are plants green? The answer is starkly simple: they reflect green light because of the pigment chlorophyll (mostly chlorophyll a) that is densely packed in leaf (and sometimes stem) tissue. Chlorophyll is very much adaptive to the plant in that this unique pigment absorbs light at the red and blue parts of the spectrum. The energy thus absorbed is used to initiate the complex process of photosynthesis, the process that supports the vast majority of life on Earth.[9] But the greenness of plants is not, in and of itself, adaptive. Plants are green merely because they do not absorb at the green wavelengths, reflecting green instead. Being green is an evolutionary by-product of natural selection.

I dwell on these points because when you really look at how ecosystems function, you soon realize how multilayered nature is. This insight underscores the importance of scale in space and time. Let's consider, for example, some blue jays in an oak forest somewhere in eastern North America and see where it leads.

Blue jays (*Cyanocitta cristata*) are common, easily recognized birds. Almost any oak forest will harbor a population of blue jays. A blue jay might not know that it is a blue jay in any profound philosophical sense, but any blue jay knows perfectly well whether it is looking at or hearing another blue jay or some other species of bird. Any blue jay that fails in this requisite species recognition is doomed to genetic failure, and thus such errors are extremely rare. So, blue jays, by and large, know each other. They establish territories, engage in courtship behavior, mate, and raise more blue jays. Sometimes they compete for nesting territory or food or both. Sometimes they form mobs, raucously squawking at an eastern screech-owl (*Otus asio*) at roost during the daytime hours, their mobbing behavior signaling the presence of a predator to other blue jays and even other prey species. Sometimes in the autumn they form large flocks, leave one forest, and migrate to another region where food is more plentiful. Such is the nature of blue jay populations. And underlying this pattern of behavior is the fact that blue jays consume lots of acorns, the seeds of the oaks. But they do so with a bit of a twist, a huge twist for the oak trees.

Acorns are seeds, not fruits. Jays peck and destroy acorns as they ingest them, so unlike the desert thrashers, jays are seed predators.

But jays don't eat all the acorns they pick. They typically collect and bury more acorns than they consume. Why might that be? Consider that acorns are not produced year-round, but only for a limited time period, typically late summer. Storing acorns, like storing canned food in a pantry, provides food for times when the trees fail to provide them. Any jay with a predilection for burying an acorn and then locating it to consume at a later date would have a better chance of living to eventually reproduce. If such a trait had a genetic basis, it could be passed on to the jay's offspring and eventually become widespread among jays, an adaptation for winter survival. And indeed, some other species of jays also store seeds.

Acorn abundance can vary remarkably from one year to the next. So the jay population eats a few acorns but buries many others, either knowingly or unknowingly (we have no idea what jays know and do not know) providing food insurance for the upcoming winter. Some, perhaps most, of the buried caches of acorns will not be reclaimed by jays, but will instead be consumed by other seed eaters, or broken down by bacteria or fungi. But a few subsequently sprout into new oak seedlings. So blue jays plant oak trees. Indeed, jays undoubtedly facilitated oak recolonization of northern regions following glacial retreat.

The jays' behavior of eating acorns, as noted earlier, makes them predators of the oaks. But at a broader scale, jays also have a mutualistic relationship with the trees. Were it not for the seed-caching behavior of the birds, the oaks would reproduce less successfully. Acorns, as seeds go, are heavy, do not blow in the wind, and cannot move one jot after falling from the parent tree. But that is a bad place for a seed to be. If a seed germinates near its parent tree, it faces immediate stiff competition from an organism, the parent tree, with almost the same genome (array of genes). That means that the parent tree requires the same resources from the environment as its offspring but more of them (it is, after all, a mature tree). So it is advantageous for seeds to be dispersed some distance from parent trees. Jays help do this for oaks, just as thrashers do for desert chili plants (though in a different way). The jays carry seeds away from the parent tree, an act that helps ensure the dispersal of the oak seeds.

Where blue jays place seeds, how they choose their caching sites, is poorly understood. However, similar jay species in western North America have been studied. What is known is that at least among the western jay species, the birds are highly selective in their choices of sites, not random, and they remember. They successfully relocate seeds months after caching them.[10] Jays, members of the crow family (Corvidae), are among the most intelligent of birds. The sophistication of corvid intelligence has been compared with that of apes.[11]

Acorns are large seeds with generally high quantities of fat and other nutritious chemicals. These are essential to the germination and initial growth of the seedling oak tree. But they also make acorns desirable food for many animal species, not just blue jays. Numerous other vertebrate species, including raccoons, striped skunks, white-tailed deer, black bears, gray foxes, gray and fox squirrels, eastern chipmunks, white-footed mice, red-headed woodpeckers, wild turkeys, and numerous insect species also share in the acorn feast. Perhaps the greatest acorn predator of them all was the now-extinct passenger pigeon, whose population once numbered in the billions. What might be the collective effect of such an array of seed predators on the combined reproductive efforts of the various oak species?

The various members of the animal community that consume acorns might in theory eat them all, and that would leave the oak trees essentially infertile, their collective reproductive success a big fat zero. Further, any forest is inhabited by a diverse community of bacteria and fungi that also require complex organic compounds, such as those found within acorns. Once an acorn is buried it might become food for bacteria or fungi. Ah, but then again, there is natural selection at work.

Chilies are far from unique in producing chemical deterrents to seed predation. Many oaks (as well as most other tree species) produce numerous chemical compounds, among which is a diverse group called tannins. Tannins have not been investigated as to how they may affect jays, but they influence behavior of gray squirrels (*Sciurus carolinensis*), which, like blue jays, frequently cache, as well as eat, acorns. These rodents bury acorns high in tannin content and immediately consume those that lack concentrated tannin. Tannin-laced acorns, typically those produced by a group of oak species collectively called "red oaks," sprout later than those with lower tannin levels, a group called the "white oaks." The squirrels have adapted to this oak characteristic by treating tannin-rich acorns as "storable" and tannin-poor acorns as "perishable." The red oak acorns are not eaten until a later date. Thus the tannin is being used as a signal by squirrels as to which acorns to eat and which to store.[12]

Every few years, oaks produce immense numbers of acorns. Trees do not possess any form of nervous system or brain. But they are sensitive to subtle environmental cues. In some years all of the oaks within a region, in synchrony, will have bumper crops of acorns. Perhaps some climatic signal influences acorn production. It is uncertain how this precise synchrony comes about, but as you will see, it is obviously an adaptation, a result of natural selection. In bumper years acorns cover the forest floor, far more acorns than any combination of jays, squirrels, turkeys, chipmunks, raccoons, bears, bugs, and microbes could possibly consume or destroy. The animals may stuff themselves, but the collective fecundity of the trees far exceeds the collective appetite of the consumer animals. For any acorn devoured perhaps thousands remain untouched (though precise figures on this point are lacking). But anyone who walks within such an oak forest in a mast year certainly steps on plenty of acorns. Many acorns are cached by jays and squirrels. By synchronizing their reproductive efforts to produce an unusually large acorn crop, the oaks have made it impossible for all the acorns to be consumed, thus ensuring that some acorns will survive and sprout. It appears that which ones do and which ones don't is largely a matter of chance but, again, no comprehensive data exist on this point.

The term "masting" is used to describe synchronous seed production in plants. Oaks are not the only masting species. Some species of pines, spruces, firs, hickories, and beech also mast. Masting has evolved independently in many plant species, and the most likely explanation for it is in response to seed predation by animals. But there could be alternative explanations for the phenomenon. For example, varying environmental conditions from one year to another could also account for masting. If, for example, the plants experienced favorable growing seasons for several consecutive years, they might store up sufficient energy to produce huge seed crops in a given year. Contrasting explanations are hard to untangle in evolutionary ecology and need not be mutually exclusive. Did seed predation select for masting or does it result entirely from environmental conditions? Note that the seed predation explanation is in large part historical, assuming a certain evolutionary history that favored the evolution of masting. As the late evolutionist Stephen Jay Gould often pointed out, evolution, to be fully understood, must be viewed in much the same way as a historian views human history.[13] Adaptations evolved in evolutionary time and thus adaptive traits we observe today are best understood when seen in historical context. It is, indeed, an "evolutionary play."

Once considered the most numerous bird on Earth, the passenger pigeon (*Ectopistes migratorius*) has been extinct since September 1, 1914, when the last bird died in a zoo in Cincinnati, Ohio. Just a century earlier the species was estimated to have numbered in the billions. Passenger pigeons were highly nomadic, the cloud-like flocks flying hundreds of miles in search of mast crops such as oaks and hickories. The massive flocks and nomadic behavior can be understood as an avian adaptation to masting. Mast represents a dense food source, thus allowing high population growth of the birds. It also is regionally distributed, so flying long distances is necessitated in order to keep a steady food supply. Though shooting pressure was extraordinarily intense on passenger pigeons because they roosted in dense flocks, the most significant factor in their extinction was forest clearance in the late 1700s and 1800s. That prevented the flocks from continuous access to mast crops.[14] As food supply diminished and hunting pressure continued, the birds succumbed. There are no equivalent species to the passenger pigeon. Nothing else has evolved such a dependency on masting, and it is quite possible that passenger pigeons acted as a strong selection pressure on tree species to evolve masting. Oak and hickory forests have repopulated most of eastern North America since farm abandonment in the latter part of the nineteenth century.

Passenger pigeons are gone. And note that without understanding that passenger pigeons really did exist in such vast numbers, any ecologist would be justified in wondering just why selection pressures favored masting. The truth today is that all of the likely seed predators don't seem to add up to enough to really threaten oak reproduction. But throw in a few billion passenger pigeons and you begin to see more clearly.

Several insects have evolved a reproductive pattern similar to masting. One is the group called periodical cicadas (which are distinct species from annual cicadas, insects that appear in small numbers every summer). Periodical cicadas emerge as adults en masse, and their combined staccato whine significantly raises the decibel level of a summer forest. They live most of their lives as nymphs (immature life cycle stages) in soil, dining on roots. They emerge synchronously in cycles of either thirteen or seventeen years (depending on species). In the summer of 2008, seventeen-year cicadas emerged from the soil around my neighborhood on Cape Cod. Dozens seemed to cover every tree branch. The landscape was a continuous cicada din, the red-eyed, chunky black insects inexhaustibly abundant. Like masting trees, cicadas "flood the market" and overwhelm the ability of predators to consume the lot. A friend asked me, "Why aren't there more birds eating these things?" Many are eaten before they can reproduce, but most are not. Birds were hopelessly outnumbered and those birds present were soon satiated. Most cicadas were not devoured and, in their short adult lives, managed to produce the next cicada generation.

Mayflies emerge synchronously from ponds in the summer months and live about a day. Their sole function during their incredibly short adulthood is reproduction. Synchronous emergence helps ensure that males find females, but it also very much reduces the overall impact of predation in a similar manner as just described for the cicadas. Schooling in fishes, flocking in birds, and herding in mammals all serve at least in part to reduce the specific probability that any particular animal will be preyed upon.

After a mast year, production of acorns may drop to next to nothing. This may be due to the oaks having utilized much of their stored energy for acorn production in the previous year. The acorn eaters go from feast to famine. In a bumper acorn year, species from blue jays to gray squirrels typically have very successful reproduction, thus boosting their regional population sizes. The lack of acorns following a masting year forces many of these animals to emigrate in search of more productive areas. In such years, large flocks of blue jays migrate by day, moving long distances in search of food. Research in oak forests in western Virginia has shown that gray squirrel, white-footed mouse, and eastern chipmunk populations vary in concert with changes in acorn production.[15] The more acorns, the more squirrels: the fewer acorns, the fewer squirrels.

Masting adds a temporal complexity to the functioning of an oak-forest ecosystem, an example of a temporal scale effect. Ecosystem processes differ in mast years. If an ecology student studies the acorn consumption and population dynamics of jays in a non-masting year, the data from that study would be distinct from those collected during either a masting year or a year when virtually no acorns were present.

Oak masting may be occurring throughout one part of the range of blue jays but not all parts. Blue jays are widely distributed throughout eastern and central North America. The concept of scale is at least as important to area (the size of the chosen study area) as it is to time frame.

Time and area scales have become recognized as essential considerations in structuring any ecological study. An ecologist wanting to study blue jay nesting behavior, for example how often the parent birds feed their young, can arguably confine the investigation to one or two study areas over perhaps one or two breeding seasons. But an ecologist who wants to learn the overall lifetime reproductive success of blue jays in a given region would have to look at woodlots of varying areas (nesting success in some bird species varies with area) for considerably longer time periods to answer such a complex question accurately.

As scales of time and area increase, the transient nature of ecosystems becomes apparent. I harp on this point because it is extremely essential that it be understood. The study of scale effects demonstrates with great clarity that nature is dynamic, always changing at various scales of space and time. Ecologists study what appear to be discrete ecosystems often giving the appearance of being in equilibrium (i.e., "balanced") but which are, in reality, small segments of a temporal and spatial continuum, snapshots of the tapestry in time of evolving life forms, a history that reaches almost as far back as the origins of the planet itself.

As time continues, so does change. Tree species are highly sensitive to temperature patterns on Earth. The current global climate change is predicted to alter the distribution of the oak forest, increasing its northern range considerably. In New England and Canada, forests of oaks may eventually replace more northern forests of sugar maples and beech. Even land that is now covered by spruce and fir trees may, in the next century, become a land of oak forests.[16]

It is a simple thing to observe a blue jay in an oak forest. It is a much more complicated thing to measure all of the activities of blue jays in an oak forest. At a certain scale of space and time the relationship between blue jays, oaks, and masting seems to become clear. But always remember, the balance of nature is a metaphor, not a reality. Gypsy moths demonstrate that pretty well.

The gypsy moth, *Lymantria dispar*, is an invasive species. Like many invasive species, the insect is not native to North America (an example of what ecologists call "alien species"). The species is considered invasive because it quickly can become abundant and then it exerts strongly negative effects on oak forests through the collective consumption of leaves. A population of gypsy moths is capable of defoliating a forest such that it resembles a forest in winter, when leaves have dropped. Should this occur several years in a row, not only will acorn production cease but the trees themselves may perish.

Gypsy moths were brought to North America in the hopes that they could be used commercially to make silk, but some escaped into the wild early in the twentieth century. With few natural predators or parasites, gypsy moths proliferated to the point where their caterpillars became serious pests. During what are called "outbreak years," immense numbers of gypsy moth caterpillars feast on oak leaves, their collective munching audible to anyone on a trail below the forest canopy, their small black feces (called "frass") falling like tiny hailstones, littering the forest floor.

But there are forces in the oak forest that work against proliferation of gypsy moths. In a manner of speaking, enemies can sometimes be friends.

The white-footed mouse, *Peromyscus leucopus*, is an abundant species in oak woodlands. C. G. Jones et al. uncovered a remarkable example involving white-footed mice that speaks to the subtle complexities that influence the health of the oak forest and, perhaps indirectly, the health of humans who live nearby.[17]

During masting, which occurs roughly every 3–5 years, the presence of huge numbers of acorns stimulates the population growth of many acorn consumers. These animals include white-tailed deer and white-footed mice. With the food stored from a mast year, the mice can produce a winter litter, something they rarely do in non-mast years. Ecologists call this a "numerical response," when a species' population increases as its food becomes more abundant.

But the mice do not confine their nibbling to just acorns. They have quite a good appetite for gypsy moth pupae. The more mice there are, the more pupae they eat. Thus the mice, who could be thought of as destroyers of oak seeds, may ultimately benefit the oaks in that they help alleviate the threat of defoliation from a population explosion of gypsy moths.

White-footed mice and white-tailed deer become more abundant in an oak forest in the years immediately following masting years (because their populations have increased from the acorn abundance). White-tailed deer are plagued by deer ticks, *Ixodes scapularis*. These ticks, in turn, host a spirochete bacterium, *Borrelia burgdorferi*, the cause of Lyme disease in humans. When deer abound, so do the ticks and so, therefore, does Lyme disease.

White-footed mice contribute to Lyme disease spread because, like deer, they are tick hosts. The juvenile *Ixodes* ticks feed on white-footed mice, so when masting makes both mice and deer more abundant, it allows the tick population to grow and thus causes a substantially higher risk of Lyme disease. Even as the white-footed mice may be protecting the oak forest from future defoliation by gypsy moths, they may be increasing the risk to humans of contracting a serious illness. None of that, of course, matters to the mice. They are just being mice. From a human vantage point, the fact that mice reduce gypsy moth threats to oak trees is far less important than the health threat posed by Lyme disease. Humans, being humans, are highly anthropocentric. Ecosystems are not.

White-footed mice do not always exert a strong predatory influence on gypsy moth pupae.[18] When acorn crops are low, the mice decline in population, regardless of the abundance of gypsy moth pupae. It is then that the gypsy moths begin to increase rapidly and defoliation of the forest can result. Eventually, the moths are so densely populated that a virus will spread through their population, causing it to crash.

So are oak forests examples of the balance of nature?

Like multiple layers of skin on an onion, ecological interactions at various temporal and spatial scales determine the dynamics of ecosystems and the evolution of the organisms themselves. Natural selection, as Darwin eloquently noted, is constantly at work. Oak forests are an ecological theater. Now you understand something of the evolutionary play.

✺ 9

Be Glad to Be an Earthling

"It's all about me" is a commonly used expression of egocentrism. Of course it's all about me. Natural selection saw to it. Why would it be about you, at least from my vantage point? You have your genes, I have mine, and mine are what I care about, more than I care about yours. Or actually, as evolutionist Richard Dawkins tells it, it's the other way around.[1] I am the product of my "selfish genes" (though, of course, they must have "cooperated" to make me) and thus my behavior (and beliefs, at least to a degree) reflects my genes' coded instructions for self-preservation. "I think, therefore I am." Sure, Descartes (who actually wrote *Dubito, ergo cogito, ergo sum*) sounded profound. I do think and I am what I think. Thinking is a huge deal. But I think very much along certain lines because I am the product of my DNA. Though the idea is anathema to many social scientists (certainly the ones I know), as an evolutionary biologist I believe I am fully programmed to care more about me than about you.[2]

The reality of egocentrism is by no means confined to *Homo sapiens*. Anyone who has studied animal behavior knows that non-human creatures are concerned about themselves, especially their personal safety. Even dumb ones get suspicious and run. But we are different. Toss in a really large and complex human brain, one that is capable of nuance and projection, and it is but a short hop from egocentrism to anthropocentrism, the notion that it's all about humans. We humans are, evolutionarily, a kind of social animal, and thus anthropocentrism is virtually preordained to occur with us. True, Earth contains far more than merely us, and we understand this perfectly well. Thus various creation myths associated with the world's multiple religions typically include other life forms, often with the belief that they were put forth fundamentally to serve the needs of us humans. The Garden of Eden account supports such a view.

There is a serious scientific side to all of this and that has to do with Earth itself, as well as the universe. In order for it to be "all about us," the planet upon which we all reside must be conducive to supporting not only life but complex multicellular, indeed, intelligent life. That is a rather tall order, as this chapter will document.

Let's begin with the universe, a vastly large place in which Earth amounts to rather less than a tiny sand grain on an inconceivably immense beach. All life as we understand it is composed of carbon atoms. Carbon is a remarkable element in that it readily forms covalent bonds (bonds where electrons are "shared") with up to four other elements or groups of elements at once and thus can assemble long chains and complex rings, the energy, structure, and information-containing molecules of life itself. Our bodies are assemblages of complex carbon compounds, of carbohydrates, lipids, proteins, and nucleic acids. DNA is a carbon-based molecule, a unique double helix, two intertwined chains of carbon, nitrogen, and phosphorus atoms so intricately and precisely arranged that the molecule carries coded instructions for making more of itself and of the organism it represents.

Carbon is forged, as are other elements, in the interior of stars. In 1954 the brilliant British astronomer Fred Hoyle[3] solved a big question about what chemists and physicists call the resonance of the carbon atom. Hoyle addressed the problem of how the universe came to contain such an abundance of carbon (and thus have the potential to evolve life). He showed theoretically that (within stars) two atoms of the element helium would fuse into a single atom of beryllium and that subsequently, with the help of a "resonance channel," an atom of beryllium would combine with an atom of helium to make carbon, an element whose nucleus is composed of six protons and, in its most common state (carbon-12), six neutrons. Hoyle's theoretical work was subsequently confirmed and the unique nature of the carbon atom became apparent. Unique indeed. Should it be the tiniest bit different, its key properties would change and life would not exist. Hmm.

And so it goes for many characteristics of the universe. From the slight asymmetries associated with the earliest moments of the big bang, to inflation theory, to numerous other dimensions (indeed, even to the dimensions themselves), our universe is uniquely suited to contain—us! Consider that our universe is thought to be only 4% composed of "ordinary matter," the elements that make up everything from stars to seaweed. Twenty-two percent of the universe is thought to be dark matter, thus far too dark to be really identified, but gravity says it's there. And fully 74% of the universe is composed of the most mysterious of all, dark energy, a thus far unexplained force that is causing the expansion of the universe to accelerate. That's a pretty odd universe, but then again, since it is the only one we know, maybe not.

Given our propensities for egocentrism and anthropocentrism, it is not surprising that in recent years scientists have increasingly focused on something they have named the *anthropic principle*.[4] It is hard for physicists and cosmologists to ignore the fact that many characteristics of the universe, should they vary ever so slightly, would eliminate the possibility of life as we currently understand it. So, to many anthropocentric humans, even smart ones like physicists, it appears as though the universe is uniquely constructed specifically to support life. That, of course, really means us.

The anthropic principle exists in two forms, the *weak anthropic principle* (WAP) and the *strong anthropic principle* (SAP). The details separating these are not essential for discussion here. Both assert that life exists because of a series of highly improbable events that together characterize the universe, each of which is necessary for life. The concept of the multiverse, where many different universes must exist, at least one of which happens to possess the unique characteristics of our universe, is part of SAP.

The anthropic principle, whether in its weak or strong iteration, is not to be taken lightly. It is based on theoretical and experimental physics and cosmology, and the questions and research surrounding it get to the very core of matter, energy, and how universes form. It is the obvious philosophical undercurrent that is undeniably associated with the anthropic principle that is dubious. The name alone, "anthropic," says it. Some use the anthropic principle to "prove" the existence of God. This view moves the focus of the discussion dangerously close to the slippery slope of teleology and purpose. The universe must have purpose. We are that purpose.

Earth too, seems compellingly "anthropic." Except that astrobiologists (those who optimistically search for life elsewhere than Earth) use a cuter and less loaded term, the "Goldilocks effect." Once upon a time a little girl named Goldilocks (who was known to associate with bears) was very selective about eating porridge. It had to be "just right," not too hot and not too cold. Astrobiologists compare Earth with the famous porridge of Goldilocks. Its various characteristics (as they relate to those of the Sun) make our planet "just right" to support life. Earth is within the "habitable zone" of our Solar System. And what might some of these habitable zone characteristics be? If you were asked to list what variables you would want to measure to establish that life exists on an unknown planet, what would you say? Begin by pouring yourself a glass of water.

Life exists on Earth because Earth has liquid water. It's almost that simple (but in science, of course, nothing is ever *that* simple). Water is key. But life also needs precursors: diverse, low-energy inorganic and organic compounds and an energy source that will, under the right conditions, allow the synthesis of energy-rich complex molecules. The good news is that low-energy organic compounds seem abundantly represented in the universe (they are contained in comets, for example), and many scientists now believe that Earth was bombarded, early in its history, by numerous objects from space, many containing organic "precursor" compounds as well as water. These chemicals were the original building blocks of what would become life. The Sun, about which more will be said shortly, supplied lots of energy, the nascent oceans filled with liquid water, and thus the synthesis of complex organic compounds was facilitated.

Water is plentiful in the universe, but not liquid water, not oceans, lakes, and rivers. Comets are mostly water, for example, but in the form of ice. Without liquid water, the chemistry of life as we understand it could not occur. Life is a continuous and complex series of organic reactions all accomplished in aqueous solution. Consider the fact that we die a lot quicker from dehydration than from starvation. Liquid water is essential, always has been. Looking at the eight other planets in the Solar System (yes, I'm counting Pluto, and I know that it has been demoted to being a "dwarf planet," also known as a "plutoid"), none support ice-free surface oceans, and such a fact reduces the likelihood of life forms (at least as we know them) on those places.

Evidence suggests that water once flowed in abundance on the surface of Mars and may have even flowed as recently as a mere few million years ago. Even now it may flow under the seasonal snow pack. Mars Orbiter photographs show clearly the channels of ancient (and perhaps not so ancient) waterways in many places on the planet.[5] Where there is liquid water there may be (or have been) life. It is worth a look. As I write this chapter NASA's Phoenix Mars Lander is scooping up soil samples from the Martian north pole to be used in experiments seeking to learn if the soil chemistry is indicative of living or once-living forms. What news that would be!

The reason why astrobiologists are immensely curious about Jupiter's icy large moons is that ice that covers their surfaces may rest upon deep liquid oceans. Maybe things are swimming in them. Europa, Ganymede, and Callisto, Jupiter's largest moons, each may have subsurface oceans. Wherever there is liquid water, it is theoretically possible that there is something in it that is metabolizing.[6]

When asked to identify the star nearest Earth, some people are tempted to reply that it is Proxima Centuri, whose distance is a mere 4.24 light-years from our planet. But that answer is incorrect. The nearest star, of course, is the Sun, much, much closer at 93 million miles (150 million km) or a mere eight light-minutes away. This distance, as Goldilocks would be quick to note, is "just right" for sustaining oceans of liquid water. But the Sun does more to permit life to exist and evolve on Earth.

The process of photosynthesis, key to sustaining life on Earth, relies on solar electromagnetic radiation (at wavelengths of visible light) as its basic energy source. The potential energy contained in the molecules produced in photosynthesis originally traveled from the Sun through space to Earth. That energy was released in nuclear fusion reactions that power the Sun as well as all of the other billions upon billions of stars in the universe. Earth is an "open system" with regard to energy input. Earth receives constant energy from the Sun (and negligible amounts from other stars). Without such uninterrupted energy, Earth would be lifeless. So for life to evolve and be sustainable on Earth, the Sun, like a good light bulb, must burn steadily and long. It does.

The Sun, as most folks (hopefully) know, is much larger than the Earth. It would take 109 Earth-sized objects to fit across the diameter of the Sun (which is 865,000 miles or 1,392,000 km), and over a million Earths could fit within the volume of the Sun. But the Sun is anything but a large star. Astronomers place it in a category called *yellow dwarf* stars. It is a very good thing for life on Earth that the Sun is in such a category.

Stars vary in mass, color, and, most importantly, in energy output. Stars burn with varying intensities, many much hotter than the Sun. Some stars are immense compared with the Sun, up to 100 times the mass of the Sun, and some are much smaller than the Sun. Some are red, some yellow, some bluish-white. Just as white-hot objects have a higher temperature than red-hot objects (a flame from an arc welder is much hotter than wood burning in a fireplace), bluish-white stars are much hotter than red stars. The Sun's surface is very hot, 6,000°C (11,000°F), but some stars are as hot as 50,000°C.

The mass of a star is critical to how fast it consumes hydrogen, its nuclear fuel. The greater a star's mass, in general, the hotter it is and the more quickly it consumes its nuclear fuel. The bright star Spica in the constellation Virgo, a "blue star," is so hot that it will exhaust its nuclear fuel within a mere 10 million years. That's not enough time to evolve life, certainly not in complex multicellular form. Should there be any Earth-like planets orbiting Spica, it is most unlikely that any will ever host living creatures.

The Sun is much less massive than Spica and falls within a category of stars whose fuel should last for something like ten billion years from inception. Given that the Sun formed about 4.6 billion years ago, it is in "middle age" right now. The margin of ten billion years of relatively steady burning is ideal not only for supporting life forms but for allowing sufficient time for complex evolutionary patterns to emerge. That means things like us.

Within the next five billion years, the Sun will begin to exhaust its hydrogen fuel and it will change to expand into a red giant star. This change will happen slowly, over many millennia. As a consequence, Earth will eventually be heated by the expanding Sun to the point where life as we understand it cannot be supported. That's real global warming. All ecosystems will cease to function, all life on the planet will cook to extinction. Perhaps the descendents of we humans will have colonized elsewhere in the Solar System or even some distant spot in the Milky Way galaxy. Science fiction will have become science fact. That, of course, depends upon our civilization surviving up to that point.[7] As for the Sun, it will, in its very old age, shrink to become a white dwarf star, with little energy output, no longer a place that supports a planetary system where there is life.

Don't worry about the Sun dying. That is a good long ways off and other problems are more pressing. But do appreciate that the Sun is about at "middle age" for the type of star that it is. That's a huge deal. The Sun (and the rest of the Solar System) coalesced (as stated earlier, about 4.6 billion years ago) from a gaseous cloud, compressed by gravity, and composed of dust containing elements forged in earlier stars. The universe formed 13.7 billion years ago. Though life on Earth likely evolved by 3.5–3.8 billion years ago, within a billion years of Earth's formation, multicellular life did not evolve until much later. It was not until about 1.8 billion years ago that complex eukaryotic ("true nucleus") cells evolved. All plants, animals, and fungi as well as protozoa are composed of eukaryotic cells. Such cells are larger and far more complex than prokaryotic cells, which make up various kinds of bacteria. According to the fossil evidence we have, it was not until just over 600 million years ago that multicellular life evolved. It was not until about 6 million years ago that anything resembling hominids evolved. Creatures walking upright with brains essentially like us

did not appear until about 200 thousand years ago at the absolute
earliest. So, though it took "only" a billion or so years to evolve
life on Earth, it required another three-plus billion years for that
life to get smart. Intelligence "don't come easy." (And bear in mind
that we humans are only about 2% distinct in our DNA from
chimps and chimps are, to speak the plain truth, pretty dumb compared with us.)

Appreciate the distinction between the probability of evolving
microbial life and that of evolving intelligent life. The former may
be widespread throughout the universe. The latter, not so much.

One of the most intriguing results of recent astronomical research
is the confirmation that stars with planetary systems are commonplace. This, of course, does not mean they are all brimming with
life, though some "Earth-like" planets have been identified. Using
the basic distribution of star types as a database, attempts have
been made to calculate the number of different habitable planets
that might exist throughout the universe. The famous Drake equation[8] is such an example. The equation multiplies a series of variables, each of which is mostly an educated guess about what is necessary for life to evolve, exist, and persist. You begin with the
number of stars in the Milky Way galaxy and multiply that by (1)
the fraction of stars with planets, (2) the number of planets in a
star's habitable zone (Goldilocks effect), (3) the fraction of habitable planets where life actually does arise (presuming life is not "inevitable"), (4) the fraction of planets where complex life forms arise
(much smaller!), and (5) the percentage of the lifetime of a planet
that is marked by the presence of complex life forms. Given that
somewhere between 200 and 400 billion stars exist in the Milky
Way galaxy, the Drake equation leads to the possibility that as
many as a million intelligent civilizations might be found throughout our home galaxy (and remember, the universe consists of billions of galaxies other than our Milky Way). But don't get your
hopes up just yet, there's more.

In a book with the intriguing title *Rare Earth: Why Complex
Life Is Uncommon in the Universe*,[9] Peter D. Ward and Donald
Brownlee make a compelling case for why the Drake equation may
vastly overestimate the abundance of complex life.

Ward and Brownlee add more variables to the equation, including the fraction of stars with metal-rich planets; the number of stars in the galactic habitable zone; the fraction of planets with a large moon; the fraction of suns with Jupiter-sized planets; and the fraction of planets with a critically low number of extinction events. Adding these variables significantly reduces the number of planets in the Milky Way likely to support sentient complex life. But why add them? What's the deal with being metal-rich, with having a big moon, with having Jupiter or something like it?

Metal-rich planets are necessary to provide essential elements that are critical to life, elements that have key roles in various metabolic processes. Atoms of iron are at the core of the hemoglobin molecule and atoms of magnesium are essential to the chlorophyll molecule. Atoms of silicon, calcium, potassium, even selenium are each essential to numerous life forms. Not all planets are so endowed. Neptune is a gas giant, for example, with a dense wind-driven atmosphere composed of 80% hydrogen, 18% helium, and about 1.5% methane. No one knows for sure, but there is thought to be a small (Earth-sized) rocky core deep within Neptune's gassy exterior. Even if it were in the habitable zone, it seems unlikely that life could evolve on Neptune.

Not every place in the galaxy is conducive to the evolution of life. The center of the Milky Way, like that of other galaxies, apparently consists of a massive and highly active black hole. That is not a good place to locate a life-sustaining planet. Indeed, recent evidence suggests that stars near the galactic center, located about 26,000 light-years from Earth, are very young and perhaps short-lived. For the evolution of life it is best to be in the galactic suburbs, as we are, located on a spiral arm, far away from the worst neighborhoods. The "inner city" of the galaxy is too dangerous.

And the Moon? Why does life depend on having a large moon? It probably doesn't at the microbial level, but, regarding the planetary stability (key word here) necessary for evolving complex multicellular life, do not thank your lucky stars, thank the Moon. Earth's only natural satellite is unusual as planetary satellites go. It is proportionally large relative to its planet, indeed, proportionally the second largest moon in the Solar System. Only Charon, Pluto's moon, is larger in proportion to its planet (but recall that Pluto is now classed as a dwarf planet). Earth's diameter is 7,926 miles (12,756 km) and the Moon's diameter is 2,160 miles (3,476 km), which means that the Moon is about 0.27, just over 25%, the diameter of the Earth. Jupiter's Ganymede is the largest satellite in the Solar System, with a diameter of 3,268 miles (5,260 km), larger than Earth's Moon. But in proportion to its parent planet, Jupiter, 88,842 miles (142,947 km) in diameter, Ganymede's diameter is a mere 3.7% of its parent planet. The mass of Earth's moon compared with Earth itself vastly exceeds the mass of Ganymede compared with the mass of Jupiter. Earth and the Moon are unusual in that astronomers characterize them as virtually a bi-planetary system, given the large proportional size of the Moon to the Earth. Thus the close proximity of the Moon means its gravity exerts a strong effect on Earth.

What if Earth had no moon? Yes, there would be no moonlit nights, many romantic songs would not have been written (my 1963 college prom had the theme "Moon River"), eclipses would not occur, the grunion would not spawn, and wolves would have nothing to howl at. But the consequences of moonlessness would be more ecologically profound.

The Moon, which today is on average 238,860 miles (384,400 km) from Earth, was considerably closer to Earth when it formed, though exactly how close is a matter of conjecture. Today the Moon is becoming more distant from Earth, receding at about three centimeters annually. But the proximity of the Moon to the Earth, and its proportionally large size, means that throughout its existence, the Moon has exerted a significant gravitational effect on its planet. Most of us realize that Earth's tidal cycles are caused mostly by the influence of the Moon (acting with the Sun, of course). Given that life may have originated in conditions prevalent in tide pools and other coastal environments, the Moon may have indirectly contributed to the first appearance of life on the planet.

What is generally less well known, and what is more important, is that the Moon stabilizes the tilt of the Earth in space, what astronomers call Earth's obliquity. If the Earth's obliquity had undergone numerous substantial changes, making the planet "wobble" unpredictably, Earth's climate would have undergone far more frequent, severe oscillations, possibly too severe to permit complex multicellular life to evolve. Mercury, Venus, and Mars have gone through what are described as "chaotic" alterations in obliquity. Neither Mercury nor Venus has a moon, and Mars has but two tiny moons, not sufficiently large to stabilize the planet. Our Moon's gravitational "calming" effect on Earth may have been of utmost importance to its future inhabitants. The Moon, a lifeless place, may have helped make life more possible on its larger neighbor.[10] And the Moon's very existence may be due to an improbable and calamitous event very early in Earth's history (another example of the anthropic principle).

When Earth formed it is likely that the new planet had no moon, at least not one comparable in size to the one we all know so well. For well into the twentieth century the origin of the Moon was a great mystery. In the early 1970s I attended a lecture by the Nobel Prize–winning chemist Harold C. Urey, who reviewed each of the three prevalent hypotheses that were then offered to account for the presence of the Moon. After carefully describing why each hypothesis was badly flawed and unlikely to be true, Dr. Urey, his tongue firmly in his cheek, concluded that the Moon could not possibly exist! Dr. Urey closed his lecture with the hope that a fourth hypothesis would come to light, one that would better account for the Moon. And it eventually did.

The Apollo program, arguably the greatest technological achievement of the twentieth century, succeeded in launching six missions that successfully landed humans on the Moon and returned them safely to Earth. On July 16, 1969, Apollo astronaut Neil Armstrong became the first human being to set foot on an extraterrestrial place, the Sea of Tranquility, on the Moon. The collective efforts of the astronaut explorers, and in particular the samples of Moon rocks returned to Earth, shed immense light on how the Moon likely formed.

Moon rocks bear a striking chemical similarity to the upper mantle and crust of the Earth, the outermost layers of the planet. As bizarre as it may sound at first, evidence gathered from studies of the geology of both Earth and Moon suggests that the Moon formed in the aftermath of an immense collision between Earth and a planet-sized body. Based on computer simulations, the colliding body is thought to have had approximately the diameter of Mars, just over half the diameter of Earth.

The "giant impact hypothesis," as it is called, asserts that only about fifty million years after Earth formed, it underwent an impact with a Mars-sized planetary body.[11] The sheer immensity of such an event can only be imagined, and then with some effort. Gravitational attraction soon began to reorganize the pulverized remains of Earth, and some of the material blown from the nascent Earth coalesced nearby to form the Moon. By about 4.3 billion years ago the Moon had cooled and the period of heavy bombardment of material remaining from the collision as well as the formation of the Solar System had about ended. Much of the Moon is really "leftover" Earth.

The impact imparted a lot of angular momentum to the Earth-Moon system. Earth was spinning faster back when the Moon first formed (i.e., a day would be far less than 24 hours), and as this momentum was very gradually lost, Earth's rotation slowed. Earth owes its spin to the impact that formed the Moon, an impact that also affected magnetic field generation and even, by virtue of its affect on the atmosphere, weather patterns. So look up in the sky on the night of a full Moon and say a big "thank you."

And while you are thanking the Moon, give a nod to Jupiter, easy enough to locate, huge and bright, as planets go. Jupiter is immense in comparison not only with Earth but with virtually all other planets in the Solar System (only Saturn comes close to equaling it). Its huge mass means that Jupiter exerts a very strong gravitational attraction on things that pass close to it, things like asteroids and comets. There are many objects in space that have trajectories that occasionally cross the orbital path of Earth. The last of the dinosaurs as well as many other life forms were victims of one of these objects at the close of the Mesozoic era, an event that I will discuss in more detail in the next chapter. There have been numerous other impacts on Earth, some major, both before and after the one that ended the Mesozoic era. But it could be far worse and, to the degree that it isn't, we probably have Jupiter to thank.

Consider what happened to Comet Shoemaker-Levy. In 1992, pulled in by Jupiter's impressive gravitational effect, it passed quite close to Jupiter and was literally torn apart into several cometary fragments. More importantly, Jupiter "captured" the comet's orbit, and two years later numerous astronomers, both professionals and amateurs, observed the fragmentary remains of Shoemaker-Levy as they crashed dramatically into the atmosphere of Jupiter.[12]

The presence of a large planet such as Jupiter reduces the number of potential catastrophic collisions of Earth with such objects as asteroids and comets (but see chapter 10). Jupiter acts as a kind of cosmic "vacuum cleaner," sweeping the inner Solar System of potentially dangerous objects. While it is true that previous devastating events on Earth have been followed by the evolution of new life forms (and we owe our own good fortune to just such an event!), if massive collisions were to be far more frequent, such calamities could prevent the formation of complex ecological communities. As Charles Darwin so aptly and repeatedly noted, evolution requires time.

Ward and Brownlee assert that extinction events must follow the model of the Goldilocks effect and not be too many or too frequent. Otherwise there is not sufficient stability for complex life to evolve. Earth must be sufficiently stable over long time periods to permit evolution to accumulate life forms, as the more of those there are, the more probable will be the eventual evolution of complex intelligent life.

What it all adds up to is that Earth has a long list of unique characteristics that make complex life possible and indeed likely here. But Earth's uniqueness in this regard suggests that elsewhere in the universe complex life may be far from common, even as simple life forms might abound. It also implies that intelligent life might be quite rare.

Other factors also contribute to Earth's suitability for life.

Most of us take for granted that Earth is a magnetic planet. Compass needles point north and thus we use a compass as a navigation aid. But what if Earth, like most other planets, did not have a magnetic field? Again, there would be consequences.

Earth is constantly being bombarded by potentially harmful radiation from space, most of it from the Sun but also cosmic rays from space. The Sun emits what is called the "solar wind," and radiation of this sort could certainly be harmful to life, even preventing the evolution of large multicellular life forms. But we exist happily along with elephants and redwood trees, so how are we protected from cosmic rays and the solar wind?

The answer is that Earth generates a strong magnetic field, called the magnetosphere. In 1958 the Van Allen radiation belts (which had been hypothesized) were confirmed by Explorer 1, the first artificial satellite launched by the United States. Like flak jackets around the planet, the belts intercept cosmic rays from space and solar wind particles from the Sun, affording a magnetic blanket of protection to Earth. The Van Allen belts result from the magnetic field of the Earth that, itself, results from the Earth having a solid inner core and liquid outer core. The complex flow of heavy metals such as iron and nickel in the outer core generates the strong magnetic field that in effect protects life on the planet from the various forms of "bullets" from space.[13]

The anthropic principle and Goldilocks effect are powerful statements of the uniqueness of Earth. Earth is an ecological planet for numerous reasons, most of which are ultimately improbable. Remember Fred Hoyle, the discoverer of resonance channels in the carbon atom? He was fond of saying that the evolution of complex life forms by natural selection was as probable as a tornado blowing through a junkyard and assembling a 747 jet aircraft. As in his view of virtually all of organic evolution, Hoyle was gloriously wrong in making such a naïve analogy. Natural selection is cumulative over many generations, not tornadic.

Think of it this way. Earth and the life it sustains may seem improbable, but so do you, and you are holding this book. Consider how many potential eggs and sperm cells there were that could have combined when you were conceived. The probability of uniting that one particular sperm cell (among billions contained in just that one fateful ejaculate) and that one particular egg cell (among the thousands that had the potential to be shed in that particular cycle) to produce you is likely a good deal smaller than the likelihood of complex life being found in numerous other places within the known universe, which is now thought to be 93 billion light-years in size. Had a different sperm won the race or a different egg been shed in that cycle, you would not be you, you would be someone else (though you would, of course, think you are you). Each of us, as individuals, is what can only be called improbable. Sure, it's all about me. Conception is easy, commonplace. But each conception is nonetheless unique. Maybe it works that way with universes too. Conception may be easy but each universe unique.[14] Perhaps, as many cosmologists are coming to believe, there exists a multiverse of many universes, among which ours is very special, at least to us.

One aspect of our good fortune is that we all likely got here as a species only because about sixty-five million years ago the dinosaurs had a really bad day. That's the topic of the next chapter.

∾ 10

Life Plays the Lottery

One obvious argument against the existence of a balance of nature, at least as such a balance implies purpose and teleology, is the reality of just plain luck. Good or bad, it doesn't matter. Especially when luck changes the world.

Luck is not a very scientific term but it comes close to one that is, the word "stochastic," meaning nondeterministic. If evolution-altering events are stochastic in nature, then patterns of extinction and speciation (at least some of them, maybe many) may have little to do with anything other than "dumb luck." This reality does not reduce the importance of natural selection. Indeed, natural selection is stimulated in such situations, as it is in such situations that new species evolve.

While I was writing this chapter the local television news reported that a convicted multiple sex offender had just won 15 million dollars in the Massachusetts State Lottery. The story, not surprisingly, focused on how intrinsically "unfair" it is for someone who has committed vile acts against society to now reap an immense reward merely by purchasing a scratch ticket. That's luck for you. Think of all the decent, deserving, law-abiding folks who bought lottery tickets in that game. And who won? The sex offender. There is, of course, no connection between one's social history, one's criminal record, and one's probability of winning a state lottery. Thanks to a small act, purchasing a lottery ticket, and unusually good luck in a fully stochastic game, Massachusetts now has a multimillionaire multiple sex offender.

The study of life's history in deep time is illuminating. I think part of the reason why ecologists came so late to the recognition of the significance of temporal scale is that they work in the present, usually on short time scales (it takes about four years to earn a Ph.D.). Also, most of their models and underlying assumptions about ecosystem structure and function were until recently based on the presumption of attaining and maintaining equilibrium within ecosystems. Stochastic models were not common in my graduate training during the 1960s, though they certainly are now. That is why I am devoting this chapter to how evolution, and, of course, ecosystems, are changed by chance events (luck).

I lecture and teach about dinosaurs, and dinosaurs fascinate.[1] The question I am asked more than any other about dinosaurs is, what killed them? The large dinosaurs that roamed Jurassic Park are long gone, a fact that has produced no shortage of suggestions as to why. A perusal of my many dinosaur books, some dating back over a half-century, shows the most frequent explanation to be "climate change." Details about just how climate changed are rarely supplied, though speculation, understandably, ranges from "too hot" to "too cold" for dinosaurs.

Extinction scenarios about dinosaurs suggest that paleontologists have occasionally let their imaginations run wild. One antiquated but popular idea was that of racial or, more accurately, species senility, the notion that evolution can only take a species so far and then it dies. Usually it is considered to have become "overspecialized." The Irish elk (*Megaloceros giganteus*) of the late Pleistocene is an example. It had an immense antler rack, and speculation was that it evolved antlers too heavy for it, too burdensome. So it went extinct. In reality it was a very large animal and its antler burden was proportionally no greater than that of smaller deer. It did go extinct, but not from antler fatigue. Also consider such species as the horseshoe crab (*Limulus polyphemus*) and the ginkgo tree (*Ginkgo biloba*). Both are so ancient that they are often described as "living fossils." So never mind about racial senility.

Other fanciful suggestions include a depletion of oxygen, exploding stars emitting deadly cosmic rays that irradiated the dinosaurs, egg-eating mammals consuming all the dinosaurs eggs, mammals outcompeting dinosaurs for food, caterpillars outcompeting dinosaurs for food, too many parasites, lousy parental care, and terminal constipation brought about by devouring newly evolved flowering plants. Even cartoonist Gary Larson weighed in. In one cartoon he featured three dinosaurs smoking cigarettes. The caption was "The real reason dinosaurs became extinct." Not satisfied with that, Larson drew another cartoon that pictured a dinosaur town meeting. The caption read: "The picture's pretty bleak, gentlemen—The world's climates are changing, the mammals are taking over, and we all have a brain about the size of a walnut."

What really killed the dinosaurs? There is no simple answer, because dinosaurs existed, indeed thrived, for 160 million years and in that vast period of time new species continuously evolved and other species went extinct. Based on the average species longevity of extant animals compared to dinosaurs, the estimate is that a given species of dinosaur might have been around for anywhere from 3 to 5 million years. If you peruse the fossil record you realize that *Stegosaurus stenops* (the easily recognized dinosaur with the big triangular plates along its back, a tail with four menacing long spikes, and a tiny head with a plum-sized brain), though common in the late Jurassic in North America, was extinct for at least 70 million years before *Tyrannosaurus rex* evolved. That's a time period longer than the entire Cenozoic era, the period from 65 million years to the present, when modern mammals have dominated the planet. Needless to say, *Stegosaurus* and *T. rex* never met.[2] So all the dinosaurs did not go extinct at once, far from it. Whatever it was that took out the *T. rexes* at the end of the Cretaceous had nothing whatsoever to do with the demise of *Stegosaurus*.

There are general patterns of extinction. There is background extinction, where a single species becomes extinct while others do not. This sort of process is relatively constant, and is, of course, compensated (though certainly not balanced) by ongoing speciation. There are minor extinction events, such as the one that ended the Jurassic period, where numbers of species become extinct approximately at the same time.

And there have been five major extinction events. They occurred in the Ordovician (439 million years ago), Devonian (364 mya), Permian (251 mya), Triassic (199–214 mya), and Cretaceous (65 mya).[3] Each resulted in major losses of species (well over 50%) and essentially redirected patterns of evolution. After any of these major extinction events, it is accurate, indeed an understatement, to say Earth's flora and fauna were depauperate. Biodiversity was greatly reduced, and it required many years, measured in hundreds of thousands to millions, for something approaching the pre-extinction level of biodiversity to be regained. But realize that from such calamitous extinction come many evolutionary opportunities for the fortunate few surviving species.

T. rex and its colleagues appear to have been victims of the most widely known mass extinction event, the "K-T event," that ended the Mesozoic era and began the Cenozoic. Had that event not occurred, the diversification of mammals and the subsequent evolution of our own species are questionable at best. Large dinosaurs were ecologically dominant, as herbivores and carnivores. Mammals evolved about 230 million years ago, in the late Triassic period, about the same time as the first dinosaurs evolved. Mammals persisted along with dinosaurs but remained small throughout the Mesozoic era. The Cretaceous extinction of all remaining non-avian dinosaurs[4] gave mammals "their big break."

It should be no surprise that dinosaurs were undergoing various episodes of extinction since they first evolved in the late Triassic. The world changed rather dramatically over the 160 million years when dinosaurs dominated terrestrial ecosystems. Plate tectonics rearranged continents, bringing about climate and sea level changes.[5] New ocean basins formed, and some continental masses became partially or totally isolated. The gradual rearrangement of landmasses was likely a double-edged sword, stimulating speciation among some dinosaurs and resulting in extinction of others.

During the mid to late Cretaceous period, flowering plants ranging from magnolias and sycamores to various early grasses evolved, and with them we see changes in dinosaur communities. Long-necked sauropods (such as *Apatosaurus*) became less numerous as dinosaurs with more effective chewing capabilities such as hadrosaurs (the "duck-billed" dinosaurs) and ceratopsians (large rhinoceros-like dinosaurs with horns and head shields, *Triceratops* being the largest and best known) became diverse and abundant.[6] In the Cenozoic, a similar pattern is revealed for large mammal diversity. By the Miocene epoch (beginning about 24 million years ago), equitable tropical climates that supported forests had given way to climates much more temperate and seasonal, and forests were reduced in area as grasslands spread. New mammal species specialized for grazing evolved, while others, adapted primarily to browsing, became extinct.[7]

A book with the intriguing title *What Bugged the Dinosaurs?*[8] suggests that evolutionary proliferation and diversification of biting insects and other parasites during the Cretaceous may have had strong negative effects on dinosaurs. Insects are vectors for serious diseases, and many ecologists are researching how insects and the emerging diseases they spread may affect evolutionary patterns among vertebrates. Dinosaurs could not have been immune to such assaults.

Some paleontologists argue that dinosaurs were in decline, their species numbers progressively lowering, as the Cretaceous drew to an end. If that indeed was the case, whatever ended the Cretaceous was a mere *coup de grace*. But many paleontologists do not believe that data support the contention that dinosaurs were in decline as the Cretaceous drew to a close. Thus the extinction event was, indeed, of major importance.

Dinosaurs aside, one thing is certain and that is that the Cretaceous mass extinction affected many animal groups other than dinosaurs. Such a pattern, where many ecologically dissimilar forms of life simultaneously experience high rates of extinction, is the very definition of a mass extinction event.

Whatever happened to end the Cretaceous also resulted in mass extinction of certain oceanic zooplankton, including many species of foraminiferans. These are one-celled amoeba-like animals that have a tiny shell of calcium carbonate and inhabit the pelagic zone of the ocean. Some still thrive today. Their tiny shells collect to form Globigerina ooze on parts of the seafloor. They feast on phytoplankton, the base of the oceanic food pyramid. They are not very much like dinosaurs. In addition, all ammonites, a diverse group of predatory marine cephalopod mollusks similar to today's chambered nautilus, became extinct, every last one of them. And, before leaving the seas, note that all plesiosaurs (reptilian "sea serpent" types) and mosasaurs (large sea-going monitor lizards) perished at the end of the Cretaceous. Something apparently blew oceanic food chains apart big time.

On land, in addition to dinosaurs, all of the pterosaurs (flying reptiles different from birds and that were not dinosaurs), some with immense wingspans (one was over forty feet), also suffered total extinction. On the other hand, groups (but not all species) such as frogs and salamanders, turtles, snakes and lizards, crocodilians, modern birds, and most mammals all passed through the extinction filter. Why did many of these survive when dinosaurs and other entire groups did not? In other words, the Cretaceous extinction event looks to have been somewhat selective. Such selectivity demands explanation.

Two prominent hypotheses to assign cause to the Cretaceous extinction have emerged. They are not mutually exclusive and, in concert or individually, would have had devastating effects on global ecosystems. Oh, and each could reoccur.

These hypotheses are (1) extreme volcanic activity inducing catastrophic global climate change and (2) the impact of an asteroid six miles (10 km) wide in the area of the Yucatan Peninsula.[9] Take note here that there is little doubt that both of these possible causes really did happen. The debate centers around the actual ecological effects of each and thus the degree to which each contributed to the mass extinction.

Looking first at volcanism, there is strong evidence for sustained and extensive volcanic activity at the end of the Cretaceous at an area known as the Deccan Traps in India. The word "Deccan" refers to "southern" in Sanskrit and the word "trap" means "staircase" in Dutch, for the steplike appearance of the lava flows. The Deccan Traps area is immense. In some places the thickness is 500 feet, and in western India there is evidence of an 8,000-foot-thick lava flow. At its height, a total of 772,000 square miles may have been covered, lava and its associated gases continuously extruding from active volcanos. The immensity of this volcanic activity would have produced dramatic global climatic effects. Volcanic ash could have largely obliterated the sun and reduced global temperatures, while at the same time the release of carbon dioxide and sulfurous gases could have altered the atmosphere, producing acid rain and other serious effects. Global ecological food webs would almost certainly have been seriously disrupted.

Dating the age of the Deccan lava flows has proven somewhat difficult, but evidence now suggests that the volcanism was in the Maastrichtian epoch, beginning a few million years before the end of the Cretaceous, so the timing seems about right.

Evidence of an asteroid impact was first published in 1980 by a team headed by Luis and Walter Alvarez, father and son. Luis is a twice–Nobel Prize–winning physicist, and his son Walter is a noted geologist. They were not investigating dinosaur extinction. Rather they were attempting to explain the origin of a unique clay layer at the Cretaceous-Tertiary (K-T) boundary. The thin line of clay is extraordinarily high in the element iridium, a rare element on Earth but common in asteroids and meteorites. The thin layer of red and green clay was soon found in other regions, always at the K-T boundary. It appeared that the "iridium spike" was global in extent. The clay was shown to contain "shocked quartz," quartz with fine lines that only forms during impacts or nuclear explosions. Also found were tektites, small, black glassy beads that form with impacts. Radiometric dating of the tektites indicates an age of 65.01 million years, precisely at the K-T boundary.

The asteroid impact theory was highly controversial when first put forth. A bit of a turf war sprang up when some paleontologists were outspokenly indignant that a physicist would have the temerity to butt into the dinosaur extinction question. But evidence mounted in favor of the impact theory, especially when the site of the immense crater was found.

In 1990 Alan Hildebrand, after dogged detective work, published the location of the crater, just off the northern tip of the Yucatan Peninsula. He named it Chicxulub. The Mayan name is taken from a nearby village, and means "tail of the devil."

The entire crater, most of which is under water, is some 125 miles across; there is now no doubt of its existence nor any doubt of its age, at the K-T boundary. The crater was made by the impact of an asteroid about six miles in diameter striking at an estimated speed of 30,000 miles per hour on an oblique angle that spewed masses of material toward North America. Within the first ten seconds of impact there would have been a thirty-mile crater that would eventually exceed one hundred miles in diameter. Such an event is difficult to image. The world would change in virtually an instant.

The impact, subsequent shock wave, and debris would have resulted in calamitous global fires, dense and protracted clouding of the atmosphere, and possibly intense acid rain. In short it would produce catastrophic disruption in food webs. Whatever the balance of nature might have been, it sure wasn't balanced after the impact.

The exact effects of the impact are not known with certainty. Here is what a team headed up by Douglas S. Robertson hypothesized: In the first minutes to hours after impact there was an intense overhead heat pulse that created lethal body temperatures for many organisms and resulted in widespread incineration. Fires were global in extent and burned for many days over widespread areas, wherever there was fuel to keep them going. So much for the forests and grasslands of the late Cretaceous. Dust, soot, and compounds of sulfur and nitrogen were released into the atmosphere, producing global cooling and cessation of photosynthesis, and perhaps making the very atmosphere itself poisonous. This effect lasted for perhaps many months.[10]

It is worth quoting one short paragraph from the Robertson et al. paper: "The worldwide, overhead, intense IR radiation was the first significant stress after the Chicxulub impact. It occurred during the first hours after the impact, prior to the atmospheric opacity that presumably led to 'nuclear winter.' This first event was stressful enough to kill all individual nonmarine macroscopic organisms except those protected in soils, underground, under rocks, or in water, in dense aquatic vegetation, or as sequestered eggs, pupae, spores, seeds, or roots." The asteroid impact and Deccan Trap volcanism have been considered by many to be analogous to what would happen to Earth's ecosystems in the event of a global thermonuclear war, and that is why the term "nuclear winter" was used to describe it.

You see why small animals such as mammals and lizards, simply by taking shelter, may have been able to avoid the worst effects better than the large dinosaurs. Crocodiles would have had some shelter in water. Most land plants ultimately survived, at least those that remained dormant as seeds and spores until conditions improved.

So it isn't just luck. But it is largely luck. Being small and having a low metabolism may have been the combination that best assured potential survival after the impact (and that would, of course, bode well for insects and other arthropods). Large metabolically active animals such as dinosaurs were, in essence, out of luck.

Oceanic food chains would suffer rapid devastation, as phytoplankton activity would be halted by the obliteration of the Sun. We see the result of rapid food chain disruption today in places like the Galápagos Islands during El Niño years. When oceanic phytoplankton productivity suddenly drops, those organisms such as marine iguanas and sea birds at the top of the food chain suffer major losses. An El Niño is utterly nothing compared with an asteroid impact.

Many questions remain about how, for example, such sensitive animals as frogs and salamanders could have escaped the effects of acid rain. Questions also remain about exactly how long climatic effects would have lasted and precisely what they might have been.

But as far as dinosaurs go (and they did "go"), they were presumably unable to find shelter when all hell broke lose on that fateful Cretaceous day. The impact, given the projected angle of entry, would have strewn fire and debris over much of North America, where vast herds of hadrosaurs and certatopsians roamed. They and their predators were likely fried pretty quickly. Dinosaurs were unable to shelter themselves, and their high food demands did not allow the meager numbers of animals that may have survived the impact event itself to find sufficient food in the aftermath.

Paleontologists have recorded what they call a fern spike in the sediments laid down immediately at the K-T boundary, in the earliest Tertiary. What typically happens after an area is devastated by volcanic ash, such as what occurred in 1980 at Mount St. Helens, is that ferns are the first to recolonize the affected area. Ferns quickly grow from very resistant spores, and thus a blanket of various fern species soon carpets the recovering region. Later, as flowering plants recover, ferns are gradually and largely replaced. This pattern is evident in spore and pollen profiles that are found in sedimentary rocks dating from the onset of the Tertiary. Ferns persisted even as other plants eventually began to proliferate. Note that an aspiring young ecologist, had he or she been doing fieldwork a generation or so after the impact, might well have thought the "climax ecosystem" of the region was a mixture of various fern species.

Ecologists wonder just how long it takes to reestablish diverse ecosystems following a catastrophic extinction event. The single largest mass extinction in Earth's history was the one that ended the Paleozoic era and closed the Permian period 251 million years ago. It is estimated that in excess of 90% of Earth's species became extinct. Recent work[11] has demonstrated that fully thirty million years passed before complex species-rich ecosystems were again present. This is recovery measured in evolutionary time, the time required to evolve new species. Initially following the extinction, so-called opportunistic species proliferated. One vertebrate, a synapsid reptile named *Lystrosaurus*, is believed to account for 90% of all terrestrial vertebrates early in the Triassic period. A time of thirty million years for full recovery is sobering.

In his thought-provoking book *Wonderful Life*, Stephen Jay Gould eloquently put forth the concept of "contingency."[12] Gould took his title from the 1946 Frank Capra motion picture starring James Stewart, *It's a Wonderful Life*. This film, which has become an icon of the holiday season, follows the life of George Bailey, trying to "see the world," but never escaping from the small town of Bedford Falls. As Bailey falls on hard times and contemplates taking his life, his guardian angel (Clarence) appears and shows him how all of the things he's done for his family and the citizens of Bedford Falls have mattered. Without him the town would be in chaos, lives would have been lost, and misery would prevail. It was Bedford Falls' good fortune to have George Bailey. Without him, things would have been different. Gould's adoption of the title is based on his belief that if somehow we could run the tape of life's evolution backwards and begin again at any point, the outcome would be different, perhaps vastly different. The difference would be due primarily to luck, to contingency.

Major extinctions are evolutionary lotteries. There are many losers but some eventual big winners. Gould believed that many groups of organisms survive extinction events due only to luck, not adaptation. He noted that our own phylum, Chordata, was present in the Cambrian period and through time managed to successfully pass through various extinction "filters." Had that not been the case, today the world's animals would all be *sans* backbone. Who knows if any of them would be able to read? Since a degree of intelligence sufficient to develop complex language, art, culture, and science has evolved in but one group of animals and only one extant species, namely us, Gould thought it quite possible that had things gone differently, intelligence might not have evolved, at least not yet.

Had the events that terminated the Mesozoic era not happened, the trajectory of mammalian evolution would have been different. Dinosaurs might have persisted indefinitely. The dinosaur paleontologist Dale Russell went so far as to have a model constructed of a sort of "dinosauroid," a bipedal, large-brained, intelligent creature with upright posture and dexterous hands that looks strikingly similar to something kept under wraps at Roswell, New Mexico.[13] But Russell, in a speculative "thought experiment," hypothesized that relatively large-brained dinosaurs such as *Troodon*, the model for his study, had they persisted, may have eventually evolved into intelligent beings like thee and me.

One of the oldest ideas in evolutionary biology is that of progressionism, the belief that evolution has innate intrinsic direction, a predetermined pathway toward progress. Gould argued persistently and persuasively against such a belief. Progressionism is really an idea that represents the equivalent of whacking a very square peg into a round hole. How do you reconcile two major biases, that God created life on Earth, and that nature enjoys a purposeful, meaningful balance, with knowledge that evolution apparently really does happen? The answer is progressionism. You take Aristotle's old *scala naturae* (the Great Chain of Being) and add a temporal component.[14] Life improves through the ages, indeed is destined to improve, culminating in us. Balance and purpose are maintained.

Contingency argues otherwise. Those who perish in a train wreck or some such calamity did nothing wrong, were not ill adapted, nor were they fated to perish. They were merely in the wrong place at the wrong time. Those who survive are no more deserving than those lost. David Raup summarized the issue well in his brief book *Extinction: Bad Genes or Bad Luck?*[15] Adaptation is no guarantee against extinction. Ecosystems have continuously experienced "train wrecks," disruption from outside agents, and occasionally these disruptions have been colossal. There have been winners and losers in such perturbations.

If you are fond of whacking square pegs into round holes, one way to view contingency is that the balance of nature exists but is constantly reinvented as ecosystems restabilize with new life forms following various disruptive events. And things eventually work out for the better. Or not. You decide.

And be aware that contingency is almost certainly in Earth's future as well as its past. Astronomers have identified 168 Near Earth Objects (NEOs, asteroids and comets), each of which has the potential to make impact with Earth within the next century.[16] The odds are long for any of them, so don't fret too much. But consider that the asteroid Apophis is on a trajectory that could, at least in theory, bring it on a collision course with Earth's surface on April 13, 2036. Should that happen, the "good news" is that the asteroid is a mere 1,150 feet (350 m) in diameter, far smaller than the K-T asteroid. Still, it's a big rock.

Like Apophis, most known NEOs are smaller than the one that ended the Cretaceous, but any of significant size would likely spoil your day and change the ecology of at least part of the planet. They are out there. Not only is nature not balanced, neither is the Solar System.

∾ 11

Why Global Climate Is Like New England Weather

When I moved to New England nearly four decades ago, the locals used to tell me, "If you don't like the weather, just wait fifteen minutes." The region is well known for its fickle weather patterns, but then again, so are other areas here and there around the globe. Take Colorado, for example. On one of my visits to Colorado Springs it was nearly 70°F and clear on Sunday, but Tuesday featured a blizzard that closed Denver airport for several days. In the scale of human existence, New England and Colorado weather is, indeed, changeable. Now, stepping back to the scale of deep time, so is Earth's climate, which is simply weather on a grander temporal scale. If our lifetimes were measured in geologic time, it could be well said of Earth that if you don't like the climate, stick around for a few thousand years and it will change. It always has and it will continue to do so. Climate change is natural. It happens from various causes, to different degrees, and with different rates, but it happens. And it's important.

Unless you live in a closed, windowless room with no input from the outside world, you are probably aware of the ongoing discussion or debate about global climate change. It is well known that the ten warmest years (average temperature) in the past thousand years were 1990, 1995, 1997, 1998, 1999, 2000, 2002, 2004, 2005, and 2006. That suggests a trend, specifically that global warming is occurring. If so, is the temperature rise due to anthropogenic causes, the liberation of greenhouse gases (principally carbon dioxide and methane)? Should we try and do something about it? Is it already too late to alter the trend? These are fair questions, the answers to which have rather profound implications.

It is worth paying attention to climate change because climate is the prime factor determining the kind of terrestrial ecosystem that occurs in any given area. Climate results from patterns of atmospheric and oceanic circulation and heat distribution. For terrestrial ecosystems, if you tell me the average annual precipitation and the average annual temperature, I can tell you what kind of ecosystem will predominate. It's pretty much that simple, just two variables. This part of ecology is not rocket science.

For example, if the mean annual precipitation is somewhere between 300 and 400 centimeters (118–157 in.) and the mean annual temperature ranges from 20° to 30°C (68°–86°F), lush, evergreen tropical rainforest will prevail. But if the mean annual precipitation is only about 100 cm, even if the mean annual temperature remains between 20° and 30°C, the ecosystem will be savanna, parklike grassland with scattered trees, such as acacias. Deserts are uniquely arid, receiving less than 50 centimeters (less than 10 in.) precipitation annually, often far less. Some deserts, usually dominated by succulent plants such as cacti, are as warm as or warmer than rainforests. Other deserts, typically dominated by shrubs such as sagebrush, periodically experience snowfalls in winter. Ecologists cleverly call these two desert types "hot deserts" and "cold deserts." Then there is Arctic tundra, the realm of the lemming, caribou, and musk ox, which is dry and cold, so cold that permafrost (soil at or below the freezing point of water [0°C or 32°F] for two or more years) endures throughout the year. Tundra typically receives about 50 cm (20 in.) of precipitation per year and has a mean annual temperature of only –10°C (14°F).

Mean annual temperature and mean annual precipitation organize terrestrial life into huge ecosystems called biomes. Examples include deciduous forest, evergreen tropical forest (moist forest and rainforest), boreal forest, tundra, hot and cold desert, savanna, and grassland. Biomes demonstrate that temperature and precipitation together act as powerful evolutionary selection pressures. Organisms evolve anatomy and physiology that enable them to cope with climate. Being a typical tropical tree such as a fig, for example, will not serve well in far northern latitudes. During the winter the tree would die from lack of moisture, the soil frozen, water unavailable. On the other hand, a polar bear will not thrive in a jungle. All of its elegant adaptations for enduring cold temperatures, for swimming in frigid waters, for stalking and capturing hapless seals, would be useless in the Amazon. It would overheat and perish. I told you this part of ecology is not rocket science. But nor is it trivial. Earth is as diverse as it is in large part because of climate variation both in space and through time. Life, in the evolutionary process of natural selection, tracks climate. Climate varies, thus life varies.

With a nod to Jimmy Buffett, consider the changes in biomes evident with "changes in latitudes."

Gatun Lake, part of the Panama Canal, is fairly near the equator, at about 9° north latitude and about 80° west longitude. The verdant tropical moist forests around Gatun Lake are home to tree sloths, ocelots, howler monkeys, various parrots, colorful butterflies, and boa constrictors. In contrast, Murray Maxwell Bay, also at 80° west longitude, is part of Baffin Island, at 70° north latitude. The 61° latitudinal separation between Gatun Lake and Murray Maxwell Bay is, to put it mildly, ecologically significant.

Suppose we fly, at relatively low altitude, so we can see the ecological characteristics of the landscape, from Panama to Baffin Bay, strictly following the 80° line of west longitude. We head due north from Panama City, flying first over the hot, steamy Panamanian forest and then crossing over the warm Caribbean Sea. Soon we are flying over Cuba, and then we pass just east of Miami, Florida, the northernmost fringe of the tropics. The nearby Florida Keys and subtropical Everglades support many species with tropical affinities, such as red mangrove and gumbo limbo trees, various colorful parrotfish, American crocodiles, and roseate spoonbills. Flying just east of Cape Canaveral, we are again over land as we reach Charleston, South Carolina. Coastal forests of broad-leaved evergreen live oaks, mixed with southern magnolias and various pines, gradually give way to complex forests of mostly deciduous broad-leaved tree species as we cross over the Appala-

chian Mountains. This vast biome, the Eastern Deciduous Forest, spreads beneath us as we continue north over Pittsburgh,
Lake Erie, and Toronto, Canada, at which point the ecology again
changes.

Broad-leaved deciduous forests eventually and unevenly give
way to conifer forests, fir and spruce, the characteristic species of
the Boreal Forest biome. Moose, the largest member of the deer
family, along with varying hare and lynx, reside here. Vast tracts of
boreal forest appear below until we reach Hudson Bay, where
polar bears pursue seals when the bay is frozen. Flying north, still
along the 80° line of west longitude, we pass Southampton Island
to the west and see Baffin Island to the east, eventually terminating
our flight at Murray Maxwell Bay, weather permitting (and it is
not always permitting). Baffin Island is north of treeline, cold, harsh
Arctic tundra, where snowy owls feed on collared lemmings, where
wolves hunt snowshoe hares, and where shaggy musk oxen huddle
in small herds against the frigid, unforgiving winds. Our flight,
spanning 61° of latitude, has taken us over tropical moist forest,
subtropical forest, deciduous forest, boreal forest, and tundra. This
pattern occurred because, with changes in latitude, it got colder
and drier.

Why does climate vary with latitude? Why are the polar regions
cold and the equatorial regions warm?

Equatorial regions receive more consistent direct sunlight throughout the year and are thus warm. Warmth stimulates evaporation of
water, which cools as it rises in the atmosphere, condensing, and
falling as precipitation, so equatorial regions are generally wet as
well as warm, ideal conditions for rainforests. In sharp contrast,
polar regions, because the Earth is tilted 23.5° on its axis (causing
seasonal change), are limited to receiving relatively direct sunlight
for only six months per year. Most of that is received at a low angle,
diminishing the effect of the solar radiation as a heat source (because the photons have to pass through more of the atmosphere,
heat is lost in transit). Consequently it is cold at the poles, too cold
to evaporate much water, so precipitation, too, is somewhat limited. It doesn't snow much at the North or South poles, but water
from the ocean freezes, so there is a lot of ice there. Of course all

that could change substantially within the next century if Earth's temperature continues to rise.

A trip across North America, from coast to coast and all at the same latitude, will also take you through various biomes, beginning with the Eastern Deciduous Forest in the East to the midwestern grasslands (now the corn and wheat belts) to mountains, deserts, and more mountains before reaching the Mediterranean-like chaparral ecosystems of the Pacific Coast. Ecosystems vary with longitude because complexities of topography influence climate. Mountain ranges, in particular, affect climate (see below), but other aspects of topography also exert influences on patterns of precipitation, making deserts, grasslands, or forests, all at the same latitude.

Because of the way climate varies with both latitude and longitude, continents such as North America include a diverse array of ecosystem types.[1]

Though it may seem necessary to travel extensively to visit various terrestrial biomes, it is surprisingly easy to see a representative number by traveling a very short distance. All you have to do is ascend a reasonably tall mountain range. Doing so will mean that you will experience climatic changes and thus changing ecology.

In 1899, while employed by the U.S. Department of Agriculture, C. Hart Merriam documented changes along an elevational gradient at the San Francisco Peaks, a mountain range near the Grand Canyon, in proximity to Flagstaff, Arizona.[2] Merriam described "life zones," distinctly different habitats found within specific elevations along the mountainside. The term "life zone" is roughly equivalent to the term "biome." Thus it is possible to experience a series of biomes by simply ascending a mountain.

The life zones of the American West can be experienced in the Rockies, the Sierra Nevada, or the Cascade mountains. The names of the life zones reflect the reality that moving several hundred feet in elevation is roughly equivalent to moving several hundred miles in latitude. By way of example, let me take you for a typical drive from lowest to highest elevation in the Central Rocky Mountains, say in Colorado.

Beginning at the lowest elevations, where conditions are desert-like, you are within the Lower Sonoran Zone, typically short grass or sagebrush shrub desert. You then ascend into the Upper Sonoran Zone, still arid, but with sufficient moisture to support small stature, widely spaced trees called piñon pines and junipers. As the zones converge they overlap, meld into one another, boundaries often not abrupt.

Next comes the Transition Zone, so named for the transition from a predominantly arid to a moister climate. Here, rather than small trees and shrubs, tall ponderosa pine forest prevails along with oaks and aspens. The biomass of the ecosystem has increased significantly as cooler temperatures and more moisture support added plant growth.

Continuing to climb, you next enter the Canadian Zone, essentially a boreal forest of spruces and firs, hardly distinct from what you would see throughout much of Canada.

Higher still you enter the Hudsonian Zone, named for Hudson Bay in Canada, which occurs at about 60°N latitude. This life zone is cold and windy, with significant winter snows. It is here that wind chill shapes trees into krummholz, a word meaning "twisted wood," describing the shrublike appearance of some trees. Krummholz results from the protective effect of snowfall. In winter, when it is cold and winds blow hard, branches blanketed by snow endure (at 0°C, 32°F) better than the branches exposed to severe wind chill. Hence the taller branches are killed while those under snow survive and grow in shrubby clumps. In some locations, the Hudsonian Zone supports bristlecone pine, one of the hardiest tree species, with some individuals living as long as 6,000 years.

Finally, there is the life zone beyond treeline, alpine tundra, cold, dry, and windswept. Growing among the scattered rocks, boulders, and snowfields are lichens, mosses, and perennial wildflowers, some of the same species of small, hardy plants that occur thousands of miles to the north, well inside the Arctic Circle.

Pausing to do some quick calculations, an altitudinal change of about 625 feet (190.5 m) is the equivalent of moving approximately 100 miles (160 km) in latitude. If you ascend a mountain that is about 10,000 feet high, you have ecologically traveled the equivalent of roughly 1,600 miles in latitude. How about that!

And then there is the rainshadow effect. I was once on a field trip with a busload of ecologists. All of us were attending the Ecological Society of America meeting, that year in Corvallis, Oregon. Our bus was taking us west to east, over the crest of the Cascade Mountains. Our first stop was in the cool, moist temperate rainforest of dramatically tall Douglas firs and western red cedar trees, with a lush understory of big-leaf maple and various shrubs. Even the liverworts that lined the fallen logs were huge. It was overcast and raining. Well, we knew they called it a "temperate rainforest," so we should not have been surprised at the moisture from the sky. In the dark, damp forest, photography was challenging, apertures open wide, shutter speed slow, as we leaned on the trees to steady our cameras (ecologists love photographing ecosystems to use in teaching). Continuing, we ascended the mountain up its western slope, the trees becoming shorter and more gnarled with increasing exposure to wind and cold. Rain continued but eased up as we crested the mountain. Temperate rainforest was replaced at the summit by small, twisted spruces and firs. Once "over the mountain" and on the eastern slope, the predominant forest was one of statuesque ponderosa pines, decidedly drier than the temperate rainforest. The ponderosa pine forest was at the same elevation where, on the western slope of the mountain, there had been rainy temperate rainforest. We ate a pleasant picnic lunch in a pine grove under partly cloudy and dry skies. Our final stop was to be at a lower elevation, sagebrush desert along the western edge of the Great Basin. But some among us had other ideas. I was asked to support a proposal to return to the Douglas fir forest, since "it had stopped raining." I thought it was a joke until I realized they meant it. Why my incredulity?

The marked difference in precipitation between the west and east slopes of the mountain is due to moisture-laden prevailing winds coming from the Pacific Ocean. As these winds encounter the tall Cascades, they are forced upward. As the air rises, it cools, condensing the moisture held within, producing rainfall in copious amounts. This precipitation supports the lush temperate rainforest.

By the time the wind has crossed over the peaks of the Cascades, it is largely depleted of moisture. It is therefore not possible to support temperate rainforest on the east slope, though there is sufficient moisture to support ponderosa pine forest. Finally, even this moisture is depleted and the air becomes so dry that only shrub desert occurs.

Such is rainshadow effect. The east slope of the Cascades is in the rainshadow which means that more arid ecosystems must characterize the east slope. The combined rainshadow effect of the Rockies to the east and the Cascades and Sierra Nevada ranges to the west is largely responsible for the Great Basin Desert.

I was not the only one who found the suggestion to return to the Doug-fir forest odd. Our guide did too. He patiently explained that in all likelihood it was still raining on the western slope. We continued to the desert and enjoyed the pungent smell of sagebrush, under clear blue skies. And hopefully some ecologists learned some ecology.

The distribution of ecosystems in North America has changed dramatically within human history, to say nothing of what preceded it. As recently as 20,000 years ago the Northern Hemisphere was largely engulfed in ice during the height of one of many episodes of "recent" glaciation. The peopling of North America (from Siberia) coincided with the melting of the massive Wisconsin Glacier.

Glaciation has a long history in Earth's deep time. Ice ages happened even before four-legged vertebrates evolved. Geological data indicate a major ice age as the cause of the mass extinction within the Ordovician period of the Paleozoic era some 400 million years ago. Before multicellular life became common there was a "snowball Earth" period approximately 700 million years ago in which much of the planet was extremely cold. Some speculate that the warming following the "snowball period" stimulated the rapid diversification and proliferation of multicellular life forms.

Continental positions also affect climate. In the early Cenozoic, during the Paleocene and Eocene epochs, the Earth was largely tropical. But in the mid-Cenozoic, when Antarctica broke off from the rest of the various Gondwana continents and drifted to the South Pole, it initiated a global cooling effect. The world became more temperate, and many closed forests were replaced by grassland and savanna, a trend that continues today.

Climate change is forced by changes in the concentration of various atmospheric gases.[3] At the present time the concentration of atmospheric oxygen is about 21%, but in the Carboniferous period of the Paleozoic era (about 300 million years ago) oxygen concentration was a whopping 35%. With an abundance of oxygen, invertebrates such as spiders, millipedes, and some insects grew to seemingly colossal sizes. Some dragonflies had wingspans approaching a meter. Some scorpions were a yard in length and estimated to weigh up to fifty pounds. The huge sizes resulted from oxygen abundance. Arthropods exchange gases passively, unlike mammals, which use a muscular pump to force gases in and out. Since arthropods depend on the passive movement (diffusion) of gases into and out of their bodies, only an atmosphere with highly concentrated oxygen will prove sufficient to diffuse oxygen deep into the bodies of such relatively large creatures. When oxygen concentration diminished, selection pressures changed to favor only smaller arthropods. Insects, arachnids, and their kin are small today for a reason. Were they much larger, they could not breathe.

High oxygen concentration in the Carboniferous also had other consequences, such as more frequent and widespread forest fires. It was, indeed, a different world back then.

Carbon dioxide concentration has varied throughout Earth's history. Carbon dioxide is a greenhouse gas, and thus Earth's temperature is affected by changes in CO_2. Today that change is largely if not entirely anthropogenically forced.

One of the most significant events in the history of humanity was the discovery of fossil fuels. Coal, petroleum, oil shale, and natural gas, each essential in today's global economy, were formed millions of years ago, especially during the Carboniferous period, when oxygen concentration skyrocketed. During that age of lush tropical forests, there was, for a time, an apparent lack of equilibrium between the rate at which carbon was fixed in photosynthesis and the rate at which decomposers and other consumers could oxidize it and release the stored carbon back to the atmosphere (where is that balance of nature?). Carbon accumulated in incompletely decomposed plant tissue. Gradually, inexorably, the carbon was stored as sedimentary rock within the crust of the Earth, the carbon compounds converted by pressure to long-chain, energy-rich molecules that form the basis for fossil fuels, in this case coal. The immense amount of energy contained within the trillions upon trillions of fossil fuel molecules was destined to make factories run, keep home warm, and power trains, planes, and automobiles millions of years in the future.

That time came in the late eighteenth century, a time called the Industrial Revolution. It began with the discovery that coal could be used to generate steam to power machines more efficiently. Factories now all relied on coal to power their machinery. The Industrial Revolution swept through Europe and was soon in North America.

By the beginning of the twentieth century it was clear that fossil fuels had changed the world. There would be no going back. With the discovery of fossil fuels, an immense energy subsidy was made available to humans, an energy subsidy that would become the catalyst to change the world more in two hundred years than it had changed since humans invented writing and agriculture, some 10,000 years earlier.

The Industrial Revolution had by-products. The burning of coal produces soot, particulate matter, mostly made of compounds of carbon, which soon began to accumulate around industrial centers. The famous study of "industrial melanism," one of the lynchpin examples of natural selection,[4] resulted from soot pollution on trees. But something else was being released as a by-product of industry. It could not be seen or smelled, but it was there. That something was carbon dioxide. Coal and other fossil fuels such as petroleum were burned in ever increasing amounts as industry and transportation expanded, and thus carbon dioxide began increasing in Earth's atmosphere. By the mid-nineteenth century the concentration of atmospheric CO_2 was about 270 ppm. Within 150 years, a mere "heartbeat" in geologic time, it would rise to 380 ppm, the present value, and predictions suggest that it will continue to increase, eventually exceeding 400 ppm.[5]

The onset of the Industrial Revolution closed a loop in a biogeo-chemical cycle that had been open since before dinosaurs evolved. Anthropogenic use of fossil fuels, which increased steadily through-out the twentieth century, replaced the CO^2 that had been slowly taken out of the rapid biogeochemical pool as long ago as 300 million years, stored in a "deep pool," out of circulation. And, with the burning of fossil fuels, it released it rapidly.

Gases such as water vapor, methane, nitrous oxide, and carbon dioxide act to block heat energy from passing easily through the atmosphere. Rather than making a quick, unimpeded transit from Earth to space, the heat is "trapped," retained within the atmo-sphere for a relatively long period. The more greenhouse gases there are, the more this effect of trapping heat is manifest. This phenomenon, the retention of heat energy by certain atmospheric gases, is termed "the greenhouse effect." It is tremendously impor-tant in mitigating rapid temperature fluctuations on Earth and it has contributed in an essential way to making the Earth a habit-able planet. Without the greenhouse effect it is doubtful that mul-ticellular life would have evolved.

Recall that in chapter 9 it was noted that Earth is an example of the "Goldilocks effect," situated at precisely the right distance from the Sun for water to exist in liquid form. One profound benefit of the presence of oceans of liquid water is that greenhouse gases, and in particular, carbon dioxide, can be absorbed into the oceans, and, in the case of CO^2, by a series of purely physical reactions, ulti-mately converted to insoluble carbonate, taken out of circulation. The importance of this reality cannot be overemphasized. Without the oceans, any buildup in carbon dioxide, such as from volcanic emissions, would not be correctable, and the atmosphere would continually increase in CO^2 concentration that, in turn, would trap more and more heat. Eventually, this process would "run away" and the amount of heat trapped would be sufficient to raise the temperature of the planet beyond that which life could endure. Such is apparently the case with the planet Venus.[6] With a surface temperature that would melt lead, Venus is most certainly devoid of life. It is noteworthy that organisms may have also contributed to ensuring life's future by utilizing large amounts of carbonate to form reefs and shells, thus mitigating the buildup of carbon dioxide in the atmosphere.

Although the oceans can and do absorb CO^2, it is clear that since the onset of the Industrial Revolution, atmospheric carbon dioxide concentration has steadily increased. This simply means that the oceanic absorption is not "in balance" with the industrial output, which should surprise only those who are addicted to the notion of balance of nature. The increase has correlated with the growing use of fossil fuel and, particularly in the latter part of the twentieth century, with increased global deforestation. (Deforestation results in removing large trees that take in and store carbon. Often the trees are burned, quickly releasing carbon dioxide. Sometimes the forests are replaced by agriculture, which absorbs far less carbon.)

Burning of fossil fuels and the burning of wood that normally accompanies deforestation release a significant amount of carbon dioxide, a process that is ongoing and that is altering the atmosphere to the degree that Earth is warming and climate changing. Of course there are some who disagree, arguing that climate change, if it is occurring, is not being forced by human activities, but rather is due to natural variations in Earth's climate.

A recent comprehensive study, using a technique called meta-analysis, disputes this claim.[7] In the meta-analysis, the authors looked at changes occurring on all continents and most oceans from 1970 to the present. The changes are almost all consistent with predictions that follow from assuming a strong warming trend. The study links the data with conclusions from the Intergovernmental Panel on Climate Change in concluding that from at least the mid-twentieth century to the present, increase in anthropogenic greenhouse gas has forced the observed global warming.

I see climate change literally in my backyard. I live on Cape Cod, in Massachusetts. One of my earliest books was *A Field Guide to Eastern Forests* (1988), in which I discussed plant and animal species typical of various forest types throughout eastern North America. On plate 14, describing Southern Hardwood Forests, I show tufted titmouse, red-bellied woodpecker, and northern mockingbird, along with Virginia opossum. Other plates illustrating

southern forests show species such as northern cardinal and Carolina wren. In 1960 these species would have been considered a rarity of Cape Cod. Not now. All of them, including the possum, are
now common on my Cape Cod property. Collectively ecologists
refer to them as "southern affinity species." Other southern affinity
species such as turkey and black vultures are also increasing their
northern ranges.

Other changes are evident. Tree swallows (*Tachycineta bicolor*)
are returning to their breeding territories an average of a week to
ten days earlier. Such observations strongly suggest that the birds
are responding to a global warming pattern. The physiological tolerances of bird species are, species by species, tightly tied, in an
evolutionary way, to average temperatures.

Ecologist Terry Root (who was one of the authors of the meta-
analysis cited above) showed that the winter distribution of many
North American bird species is strongly correlated with temperature.[8] Of 113 species examined, 60.2% were limited in their northern winter range by a particular isotherm. An isotherm is a line on
a map that connects points with the same temperature at the same
time. For example, the winter range of the eastern phoebe (*Sayor-
nis phoebe*), a species of tyrant flycatcher, is limited by the –4°C
isotherm (about 25°F) of average minimum January temperature.
Further, Root showed that the metabolic rate of birds at their northern winter boundary was approximately 2.45 times their basal
metabolic rate, a result consistent for many species. Root concluded
that the physiological demand placed on the birds by low temperature was the primary limiting factor in their winter distributions.
Root acknowledged that, when examined on a smaller, more local
scale, such effects might be masked by biotic interactions such as
competition and predation, overshadowing the importance of climate. Thus it is essential to examine large-scale patterns in avian
distribution to see climate-forced changes, as was done in the case
of the meta-analysis. The obvious influx of an ecologically diverse
array of southern affinity species into New England and beyond, an
event that has taken place in less than a half-century, is a clear example. These species are, I hasten to emphasize, not moving together
as an integrated, balanced community, but, as Root suggests, most

likely because they share similar physiological temperature tolerance. If it's getting warmer, some southern species, but by no means all, expand their ranges northward.

Many models predict the potential results of climate change. They all show major ecosystem changes as species individually respond to climatic selection pressures. Birds, once called "ecological litmus paper" by the British ornithologist James Fisher, are high-metabolism animals that fly, thus their distributions are among the first to alter as climate alters. It is therefore unsurprising that birds are among the animal groups most obviously being affected by current climate change. The American Bird Conservancy (ABC) has a website devoted to the potential effects of climate change on North American bird distribution.[9] In the course of this century, even more southern affinity bird species such as blue grosbeak and summer tanager will likely become breeding birds in Massachusetts, while other species such as dark-eyed junco and black-throated blue warbler no longer breed within the state because climate change has eliminated their favored habitats. The state bird of Massachusetts is the black-capped chickadee. Climate change models predict that it will be replaced in the present century by, of all things, the Carolina chickadee. This will happen as oak forests move progressively north, forcing a compression and reduction in northern hardwood and boreal forests.

Climate change, in a nutshell, will do the following:

It will force a significant rearrangement of spatial and temporal patterns in virtually all of the world's ecosystems. Assemblages of species at any given latitude and longitude today will not persist, but will be replaced by different assemblages. Today's typical oak-hickory forest will be replaced by a different ecosystem. Because natural selection acts on each and every species (and to be more accurate, on local populations throughout the range of a species), different species assemblages will result. The result of climate change will be Gleasonian individualistic communities on a large scale. Obvious human concerns include changes in the distribution of agricultural areas and sea level changes as they will affect coastal areas. Many coastal cities may be facing (to put it delicately) an overabundance of water. The National Oceanographic and Atmospheric Association (NOAA) forecasts increasingly severe weather for the remainder of the present century and places the blame squarely on greenhouse gas emissions.

If climate change is rapid (and it appears to be), it will present an evolutionary bottleneck to some (perhaps many) species, measured in high selection pressure and thus enhanced extinction risk. Many populations will diminish, and extinction rates will rise. Specialized species such as polar bears are at greatest risk. More generalist species such as American robins and coyotes will likely be fine. Overall, global biodiversity is expected to decline in the twenty-first century. Such a conclusion seems inevitable, due not only to climate change but to other anthropogenic causes as well.

Any notion of a balance of nature is surely naïve, given the reality of present climate change and its collective effects on global ecosystems that will become increasingly evident during this century. But the good news (if you want to call it that) is that there was no balance of nature to begin with. Somehow, that does not make me rest easy.

∾ 12

Taking It from the Top—or the Bottom

O n my thus far one and only trip to Africa, my group was observing the African megafauna at a wondrous place, the great Serengeti of Tanzania. It was late in the afternoon, nearly dinnertime, and we were driving along searching especially for large cats. What we came upon was a lone wildebeest calf, apparently separated from its mother and now abandoned, standing on a mudflat next to a lake. Few things look more pathetic than a baby herd animal whose herd is nowhere in sight. It stood alone bleating. I doubt it was happy. Soon we realized that it was, in fact, not alone. A single lioness appeared from dense shrubbery nearby and slowly began walking across the mudflat directly toward the little wildebeest. Some in my group wanted to leave, not wishing to bear witness to the spectacle that was presumably about to ensue. Others of us, me included, wanted to remain, knowing full well what nature really is and what happens in nature on a daily basis. We remained. I caught an oddly bemused look in the eyes of our two guides. Malidi and Stephen knew something that we didn't. The lioness continued toward the calf and then I noticed the cat's ample belly swaying back and forth as it walked. Pregnant? No, just full. She wasn't hungry. Lions kill to eat and eat to live, so when they are not hungry they are not interested, not even in baby wildebeests. As Stephen and Malidi smiled broadly, the lioness scarcely looked at the lonely calf, walking within a few meters of it. She continued into a clump of shrubs. And what of the wildebeest? Oh, it followed her, perhaps hoping to be adopted. They really are dumb and do need their mothers at that stage. Perhaps it thought it had found a new one. Likely the lioness would have a good breakfast.

Lions, as everyone knows, occupy the top of the Eltonian food chain (chapter 6), the apex of nature's food pyramid. One of the most common beliefs about the balance of nature is based on food chains. Early in my life I heard a defense of hunting, to whit, "If it were not for hunters, the deer, rabbits, ducks, quail, pheasants, etc. would overrun us." I was confused by this assertion, as it was clear to me that blue jays, robins, chipmunks, box turtles, and virtually everything else was never hunted and yet we were apparently not overrun by non-game species. I am not arguing here against hunting, merely the uncritical belief that hunters control prey populations. On the one hand, maybe, and on the other hand, maybe not.

Food chains, more accurately described as food webs, are real and represent complex interactions among species. Food web dynamics is a major area of ecological research and may be thought of as analogous to economic interdependencies.[1] How stable is the economy? We humans worry about economic stability, and we know perfectly well that things can and do change quickly in matters of economics. The same is true for natural food webs. Just as the collapse of a major company or bank will send deep ripples throughout the stock market that eventually find their way to the general consumer, so it is with food webs. Economics is no more balanced than nature, and in nature, as in economics, things change.

My first understanding of food web dynamics came because I opted out of attending my own high school graduation in June of 1962, to begin a two-week session at the National Audubon Society Camp in coastal Maine. I, along with my cousin Bruce Carrick, had received a camp scholarship from the Wyncote Bird Club. This venerable club, located near where we lived in suburban Philadelphia, included Bruce and me among its youngest members. I guess they thought we boys showed promise.

What I most recall from our experience at the camp was the total immersion I received in ecology. Thinking back, it was what convinced me to become an ecologist. The various instructors, each a distinguished professional in some area of natural history or ecology, talked nonstop in the field, at meals, during off-hours, about the interactions among the myriad organisms that together comprise nature. We learned to identify all manner of plants and animals, from birds to bugs, trees to barnacles, and we learned something of how they lived and interacted. We learned lots about food chains.

For instance, we visited a coastal salt marsh and identified the various salt marsh grasses and other plants as well as the animals, mostly invertebrates, which use these plants as an energy base. One of the plants was salt marsh cordgrass, *Spartina alterniflora*, an abundant grass covering much of the marsh. One of the animals common not only in the marsh but along the rocky coastal intertidal zone was a small unpretentious snail, the common periwinkle, *Littorina littorea*. We learned that the little periwinkles are the "sheep" of the marsh and intertidal zone, consuming algae and other vegetation, including the cordgrass. What we did not learn was just how much they have the capacity to consume. I learned that forty years later when I picked up the *Proceedings of the National Academy of Sciences* and saw a paper with the title "A trophic cascade regulates salt marsh primary productivity."[2]

The paper discussed salt marsh cordgrass, *Spartina alterniflora*, the same species I saw in Maine and have since observed in numerous marshes from Georgia to New England. The study focused on a periwinkle, this one a different species from *L. littorea*. It was the marsh periwinkle, *L. irrorata*. The researchers, B. S. Silliman and M. D. Bertness, performed a series of experiments that revealed why salt marsh cordgrass is so lush and ecologically productive. It's because animals ranging from the blue crab (*Callinectes sapidus*) to the diamondback terrapin (*Malaclemys terrapin*) eat lots and lots of marsh snails.

Silliman and Bertness did their studies in Virginia and Georgia. They placed marsh snails in enclosures that allowed them free access to cordgrass but excluded predatory animals from getting to the snails. The result was dramatic: in high (1,200 snails per m²) and medium (600/ m²) densities, snails ate most of the cordgrass, in some cases reducing the grassy marsh to almost bare mud flat. It is important to note that both of those densities have been shown to exist in nature, so the scientists were not inflating the potential population densities of the periwinkles. In contrast, when predatory animals were permitted free access to the snails, the cordgrass continued to flourish. Predation rate was monitored by delicately tethering snails to thin nylon cords attached to little stakes so that the snails had a generous range of movement, thus enabling them to feed, but they could not leave the study plot. The researchers merely had to count how many tethers were missing their snails to calculate predation rate.

Predators prevented the snails from reaching densities sufficiently high to seriously reduce the cordgrass biomass by their collective grazing. The various salt marsh "lions" controlled the molluskan "wildebeest." This is what ecologists call a *top-down force*. In other words, and to harken back to my childhood explanations, the snails did not "overrun" the marsh because the hunters controlled them.

The study is significant because for decades ecologists have believed that the uniquely high primary productivity of salt marshes is mostly due to the tidal cycles supplying marsh grasses with abundant nutrients, especially nitrogen. Such an example, as when a nutrient is responsible for enhanced productivity, is called a *bottom-up force*. The term "bottom-up" means that resources at the base of the food chain ultimately determine what happens from one step in the chain to another. The element nitrogen enhances the productivity of cordgrass, makes it grow better, but that fact does not fully explain why the cordgrass is so lush. Without the predators to reduce the snail population, the marsh would be far less productive. In some cases it might essentially be destroyed, converted to mud flat. The high primary productivity of salt marshes is due to crabs and terrapins and other predators of marsh snails, not just to lots of available nitrogen. Ecologists refer to such a system as a *trophic cascade*. What this example really shows is that both top-down and bottom-up forces affect salt marsh productivity.

Taken at face value, Eltonian food chains seem easy to explain. Begin with the Sun. That's where the energy comes from. What gets most of the energy? Plants do. They are the photosynthesizers, autotrophic primary producers. Since much of the energy they claim from the Sun is used to keep them alive and growing, that makes less available to things that eat plants, like wildebeests and locusts. So, because there is less energy available to sustain wildebeests and locusts than goes to sustain plants, guess what? There is less wildebeest and locust biomass than there is plant biomass, roughly 90% less (chapter 6). Big fierce animals like lions, tigers, and great white sharks (I know, you thought I was going to say "bears") have even less energy available to them, so consequently they are far less numerous than their prey animals. The pattern of declining biomass and abundance with distance from the Sun as measured in "links" on the food chain (what ecologists call "trophic levels") is a pattern largely forced by the second law of thermodynamics, the law of entropy. Heat is a by-product of chemical interactions, and heat is a stable form of energy. Energy converted to heat as organisms metabolize is not available to other organisms, so trophic levels always contain less energy the further they are from the base of the "ecological pyramid." If that was all there is to ecological energetics, it would be hardly worth discussing, but there's more.

For instance, why shouldn't the herbivores take it all? Ecologists have debated this issue for many years. The plants of the world, which can neither hide nor run, seem to maintain themselves very well, all things considered, against the onslaughts of multitudes of diverse herbivores. Do the plants fend off the herbivores, somehow keeping them at bay through the use of toxic compounds, resins, thorns, and other adaptations, or do the carnivores, as they did in the salt marsh example above, suppress the proliferation of herbivores, such that there are not enough of them to fully crop the plants? Are food webs governed from the bottom up or from the top down?

The discussion about how food webs are regulated, about why the world is "green," and about what forces actually structure food webs, has spawned much important ecological research. A short but highly influential paper published by Hairston, Smith, and Slobodkin in 1960 focused the debate.[3] Among various points, the authors argued that herbivores were not limited by their food source (the green plants) but rather by predation. Thus the herbi-

vores did not consume all of the plants because they were ultimately limited from the top of the food chain. The paper inspired
much critical comment, especially by Paul Ehrlich and L. C. Birch,
who argued that the implied balance of nature that was inherent in
the Hairston et al. paper was demonstrably false.[4] That criticism
was correct. Ecologists, at that stage in the maturation of the discipline, were seeking broad paradigms to describe nature and define
ecology, and more often than not, those paradigms were elusive.

Food webs vary in scales of time and space. They involve organisms that have different evolutionary histories, life cycle characteristics, and physiological tolerance limits. There are food webs
within food webs. Just think of the parasite community of protozoa, worms, and arthropods such as fleas and ticks in a wild population of a species such as a gray squirrel. From the parasite's view,
the squirrel is the resource base. Therefore, each parasite in the
squirrel food web is ultimately affected by how many squirrels are
taken by predators, or starve from lack of acorns when there is a
shortage. Acorns may ultimately regulate ticks.

In some food webs (by no means all), particular species are
disproportionately influential. Ecologists call these "keystone
species."

Sea otters (*Enhydra lutris*) live in marine kelp forests all along
the Pacific coast, from central California north to coastal Alaska.
Members of the weasel family (Mustelidae), sea otters are uniquely
adapted to thrive in the cold, nutrient-rich offshore waters that
typify the northeastern Pacific Ocean. With long, drooping whis-
kers, large eyes, and short muzzles, sea otters appear innocent and
inquisitive. They float on their backs, rocking with the waves and
often cracking abalones (a kind of clam) with rocks held against
their chests. Mother otters cradle babies on their bellies. As cute as
they appear to us, sea otters are major predators of several invertebrate species. As such, they are essential in maintaining the kelp
ecosystem in which they reside.

Like all mammals, sea otters, being endotherms ("warm-
blooded"), require prodigious amounts of food. They devour sea
urchins, abalones, and other marine invertebrates. Urchins are
spiny-skinned echinoderms, invertebrate animals rarely used for

food in North America, though their gonads are prized in Japan. But abalones, which are mollusks, are another story. Their large, meaty foot is a prized delicacy throughout the world. For this reason, historically some people have viewed sea otters as competitors for one of the sea's tastiest offerings.

Sea otters have dense fur providing insulation for the animals but also all too attractive to humans. Sea otters were hunted to near extinction in some parts of the Pacific coast in the late 1800s and early 1900s. Removing this keystone species dramatically altered the kelp-bed ecosystem, disrupting the trophic cascade (recall the example of the cordgrass and periwinkle). Without sea otters eating adult urchins, urchin populations burgeoned. The large urchin population consumed virtually all the kelp, removing the "trees" from the kelp forest. This reduced the diversity of an otherwise complex ecosystem of fish, invertebrates, and other marine algae, all of which depend on kelp in one way or another. As in the proverbial domino effect, the loss of the sea otter correlated with significant diversity loss in the kelp ecosystem.

The marine fisheries of the Pacific Coast were threatened with serious decline because kelp forests served as breeding grounds and nurseries for numerous fish and shellfish species, many of commercial value. A conscientious conservation program to protect and reintroduce the sea otters has restored the kelp ecosystem throughout most of the Pacific coast.

In 1998 a paper published in *Science* suggested a recurrence of the sea otter debacle.[5] Since 1990, sea otter populations along the western Aleutian Islands have dropped by 90%, from an estimated 53,000 animals in the 1970s to a mere 6,000 in 1997. The loss of this keystone species is disrupting the ecological cascade, affecting populations of numerous species, many with no direct interaction with sea otters.

The reason for the sea otter decline is not hunting by humans but by another mammal species, the orca, or killer whale (*Orcinus orca*). In this area in Alaska, orcas, which weigh from 2.5 to 5 tons, are generally pelagic (open sea) animals and normally not known to eat near-shore sea otters, which, at only 25–80 pounds, are marginal prey for such large predatory mammals. An orca eating a sea otter is rather like a wolf devouring a deer mouse. Researchers learned that perhaps as few as four orcas were responsible for much of the sea otter decline, a result that came as a surprise.

Why would certain orcas suddenly decide to eat sea otters? Such a food choice does not reflect optimal foraging. The orcas are spending a lot of energy on small prey, a marginal gain at best. One obvious reason: hunger. Orcas normally eat sea lions and harbor seals, much larger and more calorie-rich (per animal) than a sea otter. But sea lion and seal populations have declined, perhaps because fish populations upon which seals and sea lions depend for food have migrated elsewhere, or perhaps because an increase in commercial trawler fishing has depleted fish stocks to the extent that it has starved the sea lions and seals. Orcas, responding to an apparent reduction of their principal food stocks, switched to smaller, less desirable sea otters. They simply had little choice.

If it is true that human commercial fishing is ultimately responsible for orcas switching to sea otters, it is an example of how humans can unwittingly induce major alterations in ecosystem food webs. But that should not come as any great surprise.

Not all keystone species are top carnivores. Humid and warm tropical rainforests support many tree species. Figs (genus *Ficus*) are among the trees that produce large, nourishing fruits that are sought out by many large animal species from tapirs and monkeys to parrots and other colorful birds. The orangutan (*Pongo pygmaeus*), the red-haired great ape found in parts of Indonesia, depends heavily on a diet of figs and searches the forest diligently until it finds a fig-laden tree. Orangutans need large tracts of forest because fig trees are widely scattered.

Fruit trees form an essential resource in rainforests. In a study done in an Amazonian rainforest, ecologist John Terborgh learned that the large mammals and birds depend on a continuous supply of fruits such as figs, palm nuts, laurels, and a few others. Even though more than 2,000 species of plants were present in the forest, Terborgh suggested that a mere dozen plant species (including figs) were essential to maintain the assemblage of fruit-eating animals. Terborgh went so far as to suggest that without figs, the ecosystem would be at risk of collapsing.[6] Figs and a few other fruit-bearing plants are keystone species, uniquely important in the rainforest community.

Keystone species demonstrate that not all species have equal import. Some species have uniquely important influence on food web structure. Other species have far less influence. Ecologists realize that the influence of each species is somewhere on a continuum between keystone and noninfluential. The term "load-bearing species" has been coined to describe species of high importance in maintaining the function of an ecosystem. The loss of load-bearing species alters the food web, reducing the diversity of the ecosystem. Some keystone species support ecological cascades, where their influence at or near the top of the food chain is felt all the way to the producer trophic level. Other species, such as the fig trees, supply a critical resource base, while still others are fundamental in dispersing seeds, vital for plant reproduction. There are many examples of species that have major influences on ecosystems out of proportion to their modest abundances within those ecosystems.[7]

It is not always easy to determine load-bearing species. The American chestnut tree (*Castanea dentata*) was once so abundant in eastern deciduous forests that these forests were called oak-chestnut forests.[8] Chestnuts produced prodigious nut crops fed upon by numerous animals. But the American chestnut was not alone in supplying resources to a large array of consumer species. The chestnut crop was not the only nut crop. There was also an abundance of acorns from oak species as well as nuts from hickories. With the large-scale reduction of chestnuts due to an introduced fungal blight in the early part of the twentieth century, the forests gradually changed largely to oak-hickory forests. Otherwise there was little obvious change in the ecological community.

One of the first and clearest studies to demonstrate top-down effects was conducted by Robert Paine studying marine rocky intertidal ecosystems along the Pacific Northwest coast.[9] Many of the organisms that occupy these coastal ecosystems are limited by space available on rocks. The sea star *Pisaster* is a top carnivore of a food web involving ten animal species. The sea star is a generalized predator and as such, it feeds on a variety of prey species ranging from various snails to barnacles and mussels.

Experimental removal of *Pisaster* resulted in a radical alteration and simplification of the food web. Initially, barnacles proliferated, but they were eventually outcompeted for space by mussels. As the mussels proliferated, the algal species that occupied the rock faces were eliminated, as they could no longer attach to the rocks, now covered entirely by a dense population of mussels. Other organisms, dependent upon algae for food, emigrated, leaving the ecosystem radically simplified, all ultimately due to the absence of *Pisaster*, the top carnivore. The loss of a single top carnivore species was sufficient to initiate a process that altered the abundance of virtually all of the species at lower trophic levels, profoundly simplifying the ecosystem.[10]

Ecologists have theorized that the most powerful trophic cascades are those typical of the sea otter and sea star examples, where a single top carnivore exerts a uniquely powerful influence on the trophic level below it, thus creating a strong trophic cascade. In contrast, food webs with high predator diversity have been thought to potentially dampen or weaken trophic cascade effects. This is because should one or two predators become reduced in population, others would presumably increase. As a result, the food web would be more stable, less resistant to perturbation.

The argument is largely intuitive, though logical. It can be thought of as a "diversified stock portfolio" compared with a portfolio with but a single stock. In a diversified portfolio, should one or a few stocks decline, the monetary effects of such a decline may well be mitigated by increased value of other stocks. But nature is not a stock portfolio. Lots of research remains to be done. What is clear is that at least in some, perhaps many food webs, predators, either individually or collectively, do act as top-down forces.

To what degree do insectivorous birds affect herbivores such as caterpillars and other leaf-eating insects? It is common to watch wood-warblers, chickadees, flycatchers, orioles, and other birds foraging in the forest canopy, gleaning various insects from the leaves. But are these birds, through their combined foraging, actually exerting any significant ecological influence? In at least one case, they are. Researchers excluded birds from various white oak (*Quercus alba*) trees by completely covering the foliage with nets. In those trees where birds were excluded, there was both a significant increase in insect numbers and a significant increase in leaf damage caused by insects. The study showed that bird species collectively exerted a strong effect in mitigating the leaf consumption the insects.[11]

A similar type study in Japan[12] showed how two bird species affect different insect groups. Two insectivorous bird species were studied, the great tit (*Parus major*), which gleans insects such as caterpillars from leaves, and the nuthatch (*Sitta europaea*), which is a bark forager, taking insects almost exclusively from crevices in bark. Using exclosures in a deciduous oak forest, it was demonstrated that the great tits reduced the numbers of caterpillars and thus were directly responsible for decreasing leaf damage. The nuthatches did not eat caterpillars but instead ate ants, which the great tits ignored. Thus the nuthatches had no effect on reducing canopy leaf damage.

More recently[13] bats have been shown to limit insect herbivory in a Panamanian coffee plantation. Using exclosures, insect damage was more severe on coffee plants than when bats were permitted access. Thus top-down predation helps ensure your morning coffee fix.

In what they termed an "ecological meltdown," John Terborgh et al.[14] observed top-down effects in a tropical dry forest. The research team, working in Venezuela, took advantage of the creation of numerous "islands" of various sizes when a hydroelectric impoundment flooded parts of an extensive forest, leaving only the peaks of hills, forest fragments, exposed as newly formed "islands." Six small islands, each 0.25 to 0.9 hectares, were compared with four medium-sized islands, each 4–12 hectares, and two large islands, each more than 150 hectares (a hectare is 0.01 square kilometers or about 2.5 acres). In addition, all of the islands were compared with two sites on the mainland. The small and medium-sized islands were insufficient in area to support the normal array of predator species, but the various top carnivores all remained present on the large islands and on the mainland.

The results of the fragmentation were dramatic. The small and medium predator-free islands each experienced dramatic increases in various herbivore species. For example, the number of rodents captured on the small and medium-sized islands was 35 times greater than on the large island and mainland study sites. Iguanas (large herbivorous lizards) were estimated to be ten times more abundant on the small and medium-sized islands. Leaf-cutter ants increased markedly, especially on the small islands. Most remarkably, howler monkeys were estimated to have increased by fully two orders of magnitude, to densities approximating 1,000 per square kilometer. Normal densities are between 20 and 40 per square kilometer.

These results, considered collectively, represent obvious evidence of top-down effects. Without the predators, herbivore populations were unchecked, and the ecosystem began to radically change. Unsurprisingly, the research team found strong evidence of the increased herbivore effects on vegetation. On the small and medium-sized islands, recruitment of canopy species of trees, as determined by counting seedlings and saplings, soon appeared to be dramatically reduced. The herbivores were eating them.

The absence of carnivores should not be interpreted to mean that the ecosystems of the small and medium islands would become totally destroyed with time. They will not be. For example, howler monkey reproduction is reduced when these primates are densely populated, a form of bottom-up regulation from within the population itself, forced by its food sources, the leafy vegetation. The once species-rich forest will change, losing numerous species in the process, until, as the authors state, "the species composition of the vegetation adjusts to impose regulation from the bottom up." In other words, those plant species that are most difficult to ingest or digest, those with the most anti-herbivore adaptations, will come to dominate in the ecosystem, while other, less adapted plant species will die off. The food web will have become greatly simplified as biodiversity is lost as part of the "ecological meltdown." Think about it. It's just basic natural selection in action.

What of bottom-up effects? Bottom-up effects are most clearly seen for primary producers, the plant community. These organisms are strongly affected by factors such as amount of solar radiation, moisture availability, mineral composition of the substrate, temperature, and other abiotic factors. Indeed, recall that the factors just cited are primary determinants of biome identity, perhaps the most extreme form of bottom-up effect. Manipulation of these abiotic factors usually alters an ecosystem. Competition among plant species can be influenced by abiotic factors that may act to favor certain species at the expense of others.

Factors that maintain high plant species richness may act as bottom-up effects on the entire ecosystem. For example, in an experiment in which researchers manipulated the species richness of large aquatic plants collectively called macrophytes (translation: big plants), it was learned that greater species richness of large plants resulted in higher overall biomass. In other words, diversity promoted productivity (as reflected in greater biomass). But what also was learned is that macrophyte diversity also promoted greater biomass of algae, as well as greater phosphorus retention in the ecosystem.[15] High macrophyte species richness enhanced the functioning of higher trophic levels by providing a greater diversity of food sources, more overall cover, and better nutrient retention. As the researchers noted, "Our results imply that higher vascular plant species richness in wetlands may potentially yield up to 25% more algal biomass, thereby potentially supporting a greater abundance of fish and wildlife, and retaining up to 30% more potentially polluting nutrients, such as phosphorus." The implication of the study is that active management practices that maintain high species richness of macrophytic aquatic vegetation result in a bottom-up overall enhancement of diversity throughout the food web.

Bottom-up effects would be expected in ecosystems where herbivore pressure is most intense. Another way of saying this is that herbivores would act as a strong selection pressure on plants to evolve mechanisms to resist herbivory. One likely ecosystem in which such would be the case is lowland tropical rainforest, where the growing season is often year-round and where herbivores, mostly insects, are a constant component of the ecosystem.

Defense compounds as well as high fiber content, both characteristics of leaves that act to discourage herbivores, are proportionally more abundant in tropical vegetation. One study[16] showed a latitudinal correlation such that alkaloids (one form of plant defense compound) decreased with latitude. In other words, alkaloids are more diverse within tropical plants than in temperate plants. The reason why leaf-cutter ants (*Atta spp.*) are as abundant as they are in neotropical rainforests is that they feed exclusively on fungi that they carefully tend in subterranean gardens (hence the alternate name of fungus-garden ants). The fungi are fed leaves that are clipped by the ants and transported by them to their underground garden within a huge ant colony. Defense compounds in the leaves inhibit herbivores but they do not inhibit fungi. The fungi grow on the leaves and the ants dine on the fungi. Thus the ants avoid direct exposure to anti-herbivore compounds that numerous other insects would find totally unpalatable, and the ant population thrives. Ants, by exclusively feeding on a unique species of fungus, have for the evolutionary moment circumvented the defenses of many plant species. So the leaf-cutter ants are doing very well.

Many infectious diseases, perhaps up to 60%, contracted by humans, are initially transmitted to humans from animals. Diseases such as hantavirus pulmonary syndrome (HPS), tick-borne encephalitis, Lyme disease, and, more recently, severe acute respiratory syndrome (SARS), are known as zoonotic diseases. There is presently much concern about a virulent form of influenza present in wild and domestic bird populations that can infect humans who are in close contact with birds. In other words, in zoonotic diseases, the disease agents are found under normal circumstances in animal populations. They somehow "jump" from the animal reservoirs to humans.

Top-down control of potential disease reservoir species, particularly rodents, may be essential to preventing outbreaks of serious consequence to humans.[17] The logic is straightforward. If human activities reduce predators that exert top-down influences (typical carnivores such as weasels, cats, birds of prey), then rodents, the typical prey, will increase. Such an increase soon releases the pathogen from a form of top-down control by the predators of rodents. When rodent populations are small, the transmission of the pathogen is reduced (because it is more difficult to transmit when there are fewer animals that harbor the pathogen) and thus its numbers remain relatively in check. But as a rodent population increases, crowding permits easier animal-to-animal transmission of the pathogen, which begins to proliferate greatly. Soon the pathogen moves as an epidemic through the rodent population and becomes present in sufficient density to infect a human population. Thus top-down control of the rodent population is indirectly also controlling the pathogen population and helping avoid its spread to humans. This model is applicable to classic diseases such as black plague, caused by the bacterium *Yersinia pestis* and vectored by the Norway rat (*Rattus norvegicus*).

Food webs differ, though there are patterns. Just as no two species are identical, so it is that no two food webs are alike, nor do they respond alike to various perturbations. It is not correct to assume that more diverse food webs are intrinsically more stable than less diverse food webs.[18] Food webs experience flux as species vary in abundance or as new species enter an ecosystem and others become extinct. The serious global problem of invasive species is generally an example of how a single species that enters an ecosystem may proliferate so rapidly as to significantly alter the food web, usually by reducing biodiversity. In short, humans must learn to manage food webs to maintain whatever characteristics of them are deemed to be desirable. An example is the reintroduction of wolves into areas such as Yellowstone National Park. Wolves immediately exerted classic top-down effects on elk, reducing the elk herd and, in reality, making it more vigorous, as factors such as winter starvation and spread of disease were less likely in smaller populations. In turn, the elk asserted less of an effect on browse species such as willow, cottonwood, and aspen, which were becoming seriously reduced by what was deemed overgrazing.[19]

Food webs are intricate, but they are dynamic, changing, vulnerable, and do not represent a balance of nature. Food webs are frequently altered, simplified, etcetera. What we do to alter them or to maintain them is nonetheless of great importance. Food webs are a measure of the biodiversity of an ecosystem, and in the penultimate chapter I will discuss issues surrounding biodiversity.

∾ 13

For the Love of Biodiversity (and Stable Ecosystems?)

O ne fine June night some years ago, I was with a group of friends returning from an outing on the tropical island of Trinidad. Stars twinkled above the Arima Valley and the warm, humid night air was filled with sounds of frogs and insects. As we turned on to the long drive that led to our lodge, the Asa Wright Centre, I saw in front of us a very formidable serpent beginning a slow and deliberate crossing of the road. I abruptly stopped the van and my friends and I hurried out. In the light of the van's headlamps, and from a very respectful distance, we looked in awe and with some fear at a nine-foot-long bushmaster, a pit viper, one of the most venomous snakes in the world. The lengthy reptile didn't seem to give a damn that we were there, but we sure paid attention to it. Adrenaline levels in the mammals likely spiked well above that of the reptile. We were close to our lodge and knew perfectly well that the very trails we walked daily were part of this creature's home range. To speak the plain truth, bushmasters were reportedly "common" in the Arima Valley, a fact we understood but which was really driven home by our encounter. A close look at such an animal does give one pause. The scientific name *Lachesis muta* translates to "silent fate." Because of its great length, the snake has a long striking range and is alleged to, on occasion, strike without any warning (unlike a rattlesnake, whose audible vibrating tail signals that the animal is agitated). Long hypodermic fangs deliver a whopping dose of venom. Among its victims, fatalities are rather the norm, and thus close proximity to the serpent is best avoided. We stood and watched as the wide triangular head of the snake probed vegetation, deliberately moving into forest while the remainder of its thick diamond-patterned body still stretched fully across the narrow road. Within minutes it was gone, back into its rainforest, but certainly not forgotten.

My friends and I excitedly "talked snake" while allowing our heart rates to return to something approaching normal. All were deeply impressed at having seen such a predator. The snake had earned our respect. And no, none of us had any inclination to want the animal destroyed, merely because it was in close proximity to our lodge. Instead, we were thrilled to have seen it. Were we reacting like elitists, overjoyed by a close encounter with an animal that could kill any of us? I would prefer to believe we were enlightened environmentalists. In some intangible way, we treasured having seen this unique animal under its terms, in its forest.

Alaska brown bears (*Ursus arctos*) occasionally kill people. In 2003 Timothy Treadwell and his girlfriend Amie Huguenard were both killed by a brown bear in Alaska.[1] This event was particularly noteworthy as Treadwell was well known and widely publicized for his fondness for brown bears, often approaching them closely and even talking to them. In numerous press and media interviews he repeatedly insisted that brown bears are safe to approach if one does not act aggressively. Sadly, this well-intentioned but naïve behavior proved fatal.

Brown bears are not teddy bears. Vertebrate predators, the "big fierce animals," are best not cuddled. But then again, neither are most other animals, not to mention plants, fungi, protozoa, algae, or microbes. Biodiversity, like rocks, soil, glaciers, sand, the atmosphere, and the presence of a large amount of liquid water, all characterize planet Earth. Biodiversity is a fact, the current expression of life on Earth. Biodiversity evolved in its many forms as it will continue to do for as long as Earth is habitable. We are part of that biodiversity. But how should we regard it? What are our responsibilities toward it? What does biodiversity do for us? Why be concerned about it, even as evidence shows it to be in global decline?

Ecologists wrestle with these questions. In some cases the answer seems fairly easy because of emotional tugs associated with various species. Treadwell loved his brown bears. The majority of people I know blame Treadwell for poor judgment, not the bear that killed him. The World Wildlife Fund successfully championed the giant panda (*Ailurpoda melanoleuca*), which most people find adorable, now the WWF symbolic animal. The Chinese people mourned the loss of Mao Mao, a giant panda killed in the tragic earthquake of 2008. Whales, too, are popular. Thousands enjoy going on whale watches, admiring the huge mammals from relatively close range. Save the whales! Polar bears (*Ursus maritimus*) are currently in ecological difficulty because of melting ice in the Arctic. In spring of 2008, the federal government listed the polar bear as a threatened species. Most of my friends and colleagues strongly support measures to ensure the future of polar bears, even if they never have the thrill of seeing one in the wild. It is easy to garner public support for bears, be they brown, panda, or polar. Whales too. They make good stuffed animal toys, are featured on numerous animal documentary films, and are symbolic of animal nobility, the "call of the wild." But what of all of the multitudes of less glamorous life forms? What of deep oceanic fish that no one ever sees but which are down there nonetheless? What of army ants in the tropics and lemmings in the Arctic? What of dung beetles? Ever seen someone wearing a "Save the Dung Beetle" T-shirt? (Participants on my Trinidad trip did present me with a Bushmaster T-shirt that I wore proudly. As for the dung beetle, don't rule it out. Some ecology grad student somewhere is wearing a dung beetle T-shirt, I'd bet on it.)

Public support, such as it is, for biodiversity thus far has been much more emotionally than scientifically based. There is not necessarily anything wrong with such a justification. By our showing compassion for large animals, especially top carnivores, entire ecosystems are saved, a kind of top-down approach to conservation. But ecologists have the responsibility for researching the actual science of biodiversity. What have they accomplished thus far? We might begin by asking, just what *is* biodiversity?

Biodiversity is a word that, in the public mind, often lacks precise definition. Most people think of numbers of species when they think of biodiversity, but biodiversity can also be measured at the level of subspecies or genetically distinct population. The United States Endangered Species Act[2] recognizes the importance of subspecies in conferring protection to endangered populations such as the northern subspecies of spotted owl (*Strix occidentalis occidentalis*).

Biodiversity can also be measured as genetic diversity within a species. Doing so has strong conservation implications. For example, the overall genetic diversity of cheetahs (*Acinonyx jubatus*) is low in comparison with that of other large cat species. That makes the future of cheetah populations of concern to conservation biologists, as they have less adaptive potential due to lower genetic variability. It has been reported that cheetahs demonstrate elevated sperm abnormalities, greater susceptibility to disease, and high levels of infant mortality and infertility, all characteristics of inbreeding depression due to lack of genetic variability. But others have contested these data and suggest that concerns about lack of genetic variability are less important than habitat conservation.[3] The "cheetah debate" is an example of how conservation biology is struggling with its own message, to say nothing of its science.

Biodiversity may also be considered at the level of whole ecosystems. Old growth forest, for example, is a kind of ecosystem in decline throughout the world. Most forests are regularly cut for timber and thus do not represent old growth even though their species composition may be similar to that found in old growth. High-diversity coral reef ecosystems are also threatened in many areas due to overfishing, pollution from increased sedimentation, loss of neighboring mangrove forests, and physical abuse from tourists damaging reefs.

Just how many species inhabit Earth today? We don't know because we have not investigated the question adequately. Research on taxonomy and classification has lagged for numerous groups of organisms, and consequently, for some groups including those as obvious as beetles or as cryptic as tropical mites, we can offer only an educated guess as to their actual species number. While the figure of between three and five million extant species is often cited as the total global biodiversity, some studies have suggested that the number of species of tropical arthropods alone may be as high as thirty million[4] and that the actual number of species on Earth may approach one hundred million.[5] Robert M. May, in a comprehensive review of known species numbers for all taxonomic groups,[6] concluded that no current figure as to the presumed total number of extant species on Earth could be accepted with confidence. A more recent review[7] cited the number of described species as approximately 1,750,000 but estimated that the real total, once all species are described, would be about 14,000,000. The authors made the point of stating that their estimates are highly conservative and that the actual number could be greater. This total underestimates, since the biological species concept itself is difficult to apply to microbes and even to plants. We simply don't know how many species there are, though obviously the species richness of some groups, such as birds and mammals, is known with far greater precision than that of others, such as nematodes and fungi.

There is general agreement among ecologists that anthropogenic activity resulting in loss of habitat is the major cause of the ongoing global decline in biodiversity. But what is the actual evidence for biodiversity decline? A recently published list compiled all of the vertebrate species extinct or thought to be extinct dating from the sixteenth century to the present day, a period of about 500 years.[8] This time period represents a great expansion in human population and technology. Some of the species are relatively well known, such as the dodo (*Raphus cucullatus*) and the Steller's sea cow (*Hydrodamalis gigas*). But many are utterly unknown to the average person, such as large sloth lemur (*Palaeopropithecus ingens*) from Madagascar and the Jamaican giant galliwasp (*Celestus occiduus*), a kind of lizard. Many are island species, which tend to be far more vulnerable to invasive species and human disturbance than mainland populations. The list totals 87 mammal species, 128 bird species (including 13 species within the Drepanididae, the Hawaiian honeycreepers), 20 reptile species, 5 amphibian species (almost certainly an underestimate), and 171 fish species, all inhabiting freshwater. The list for fish includes an astonishing 102 species from Lake Victoria alone, all of which are thought to have perished from the mid to late twentieth century. The total number of recently extinct vertebrates is thus estimated to be 424 within a period of about 500 years. But most of these extinctions have occurred from the late eighteenth century to the present, so the extinction rate is actually increasing. The number of threatened, critical, or endangered species, none of which are on the list, rises annually. Vertebrates, though obvious, represent but one taxonomic group. Extinctions of insects and other invertebrates as well as most plant groups are poorly documented, but there can be little doubt that they have occurred and continue to occur.

The main factor in biodiversity decline is loss or severe fragmentation of habitat. Tropical rainforest has declined globally by over 50% of its historic coverage, a decline that continues. Given that numerous rainforest species are endemic and occupy limited areas, calculations suggest that as many as 4,000 may become extinct annually. Deforestation is high throughout all equatorial areas, and not just rainforest. In parts of South America there has been a rapid loss of other ecosystem types, principally dry savanna (called *cerrado*) and dry shrubby desert (called *chaco*). These ecosystems, once rich in endemic species, have been largely converted to soybean fields.

Though ongoing loss of ecosystems such as forests has been suggested as a primary cause of species extinctions, documentation for that assertion has been limited mostly to anecdotal observations. However, a detailed study of species extinctions following deforestation in Singapore[9] makes it clear that severe deforestation does, indeed, cause extinctions among a wide variety of taxonomic groups. Singapore, a tiny country at the southernmost tip of the Malay Peninsula, is in the humid tropics and was originally heavily forested. In the Singapore study, habitat loss is estimated to be up to 95% over a period of 183 years, from the time when the British first established a presence there. Forest reserves now occupy a mere 0.25% of the island's total area of 540 square kilometers, but they hold as astonishing 50% of the biodiversity still extant on Singapore. Major extinctions have been recorded for all vertebrate groups as well as invertebrates such as butterflies. The highest percentages of extinctions recorded were among butterflies, birds, fish, and mammals. The observed loss of biodiversity from Singapore over the 183-year period was 881 animal species (28%) out of a total of 3,196 species. These numbers will continue to escalate with any further loss of protected reserves. The authors calculated the total percentages likely to become extinct with loss of reserves by adding the number of species already extinct and the number of species restricted to reserves and dividing this total by the original number of species. The percentages are sobering in that 78% of amphibians, 39% of birds, 69% of mammals, and 77% of butterflies would become extinct with loss of what tiny forest reserves still remain. Using a species-area model (which robustly predicts the loss of species in relation to area loss), the authors expanded their study to estimate extinction rates throughout southeast Asia if current rates of habitat loss continue. Their estimate predicted a loss of between 13% and 42% of species from regional populations by the year 2100.

Other factors cause biodiversity loss. In particular the influx of invasive species of plants and animals into numerous ecosystems is resulting in reduction and loss of native species. Invasive species are so regarded because they are relatively free of predators or pathogens and undergo a kind of "ecological release" when they colonize. Some outcompete native species and others are serious predators or herbivores. For example, on the South Pacific island of Guam, the brown treesnake (*Bioga irregularis*) was accidentally introduced sometime early in the 1950s. Within two decades, the snake had increased to occupy the entire island. It lacked any predators or competitors, but, by itself, it proved to be a severe predator. Because of the snake, nine of the eleven species of forest birds have been extirpated from Guam.[10] The snake has also severely reduced other bird populations as well as native mammals. The Guam ecosystem suffered a meltdown because of a single invasive species. Numerous other examples of how invasive species reduce biodiversity, both on islands and continentally, abound and increase daily.

Human population growth in western equatorial Africa has brought apes to the verge of extinction.[11] Both gorilla and common chimpanzee populations have declined in historic times due to human exploitation in East and West Africa, but until recently these species have been safer in western equatorial Africa, in countries such as Gabon and Republic of Congo. However, ape populations in Gabon decreased by over 50% from 1983 to 2000. Human hunting pressure associated with increased mechanized logging of the region was a principal cause of the decline. In addition, Ebola hemorrhagic fever, which is nearly always fatal, is spreading among ape populations in the region. This combination of threats puts ape populations in a very precarious situation.

The continent of Africa represents a land of rapidly increasing human population and uniquely rich biodiversity. As the plight of the apes illustrates, conservation of biodiversity will be a significant challenge in Africa's future. A study of the distribution of biodiversity and people in sub-Saharan Africa demonstrated that the areas with the highest intrinsic biodiversity are also the areas most populated by people.[12] This is because primary productivity is highest in these areas, making the regions desirable for people's uses but also rich in wildlife. Thus a conflict is bound to arise as humanity wishes to expand its claims on the land. Researchers looked at the distribution of 940 mammal species, 1,921 bird species, 406 snake species, and 618 amphibian species, many of which are unique to particular areas, some of which, due to habitat loss, are threatened species. The pessimistic conclusion was that conflict is bound to arise generated by human pressures to develop the land.

The reduction of biodiversity suggests that current extinction rate is well above "background level." There have been times when speciation rate has outpaced extinction rate, as happened in the early Cenozoic with the expansion of mammalian, avian, vascular plant, and insect diversity. The world's leading expert on biodiversity, Edward O. Wilson, estimates that about 27,000 species are doomed to extinction each year.[13] That amounts to 74 per day and 3 per hour. If true, such a rate would qualify as a major extinction event. Wilson's logic for such a claim rests on numerous case stud-

ies that demonstrate the cumulative negative impacts of anthropogenic activities. For example, Wilson cites data on freshwater fish
species whose ranges occur in Canada, the United States, and
Mexico. Of a total of 1,033 species, 27 became extinct within the
past century and another 265 are vulnerable. The threats are as
follows: destruction of physical habitat (73% of species), displace-
ment by introduced species (68%), alteration of habitat by chemical pollutants (38%), hybridization with other species and subspecies (38%), and overharvesting (15%). The reason why percentages
add to well over 100% is that many species face multiple threats.

But who cares? Why is biodiversity loss of concern, other than
the esthetics of losing an irreplaceable species? Rest assured, I am
not trivializing the importance of esthetics or the ethics that may be
applied to a species' intrinsic right to survive (see next chapter for
more on this critical argument about biodiversity). But that said,
do ecologists have any substantive data on why biodiversity is actually important to ecosystem function? Simply saying we should
not reduce or disrupt the balance of nature will not suffice.

My Ph.D. thesis was devoted to correlating bird species diversity
with habitat characteristics. I was one of many ecologists investigating a hypothesized relationship between species diversity and
ecosystem stability. The idea seemed to have heuristic value. The
more species in an ecosystem, the more interactions, the more redundancy, the more stability, the more resistance to perturbation.
The concept, to repeat an analogy made in the previous chapter,
was kind of like viewing an ecosystem as a diversified stock portfolio. One stock might drop, but others would gain and one's in-
vestments would remain stable.

My work[14] showed a positive correlation between species diversity and what I deemed (by argument) to be stability, but no causal
relationship was apparent, as has been the case with many other
studies then and since. But that is changing. Today many ecologists
continue to investigate the hypothesized linkage between biodiversity and ecosystem function (particularly stability and redundancy,
but also productivity and recycling). A conference was held in
Paris in December 2000, entitled "Biodiversity and Ecosystem
Functioning: Synthesis and Perspectives."[15] Prior to the confer-

ence, various researchers were queried to graphically depict what each believed to be the relationship between biodiversity and ecosystem processes.[16] What resulted was described as a "wonderful breadth of ideas," which reflected about fifty different hypotheses about how loss of biodiversity would affect ecosystem functioning. Since the conference, additional studies have only added to the breadth.[17]

There are several ways to consider the problem. An oft-cited analogy is the "rivet analogy."[18] If, on an aircraft, a rivet is randomly removed at each flight, it is only a matter of time before the aircraft fails. The question is, how many rivets can the airplane lose before it falls out of the sky? Some rivets, such as those on the wings, might be more essential than others. The plane could likely withstand many lost rivets but a point would come, sooner or later, depending on the locations of the rivets removed, when the plane would fail. Likewise, how many species can be lost from an ecosystem before its functioning is altered or impaired? The rivet analogy in its simplest form assumes that species are fundamentally equal in their contribution to ecosystem functioning (i.e., "holding up the aircraft"). But some rivets are more essential than others just as some species have greater influence on ecosystem structure than others (chapter 12). Learning the degree to which every species affects its ecosystem is not easy.

You will also recall that keystone species such as the sea otter exert a unique influence on the functioning of certain ecosystems (chapter 12). There may be many species in an ecosystem, but only a few keystone species. Unless it is known exactly which species are the keystones or "load-bearing species," the role of biodiversity in maintaining ecosystem stability is muddled. Suppose you were to randomly remove ten species from an ecosystem of fifty species. If none of the species you removed was a load-bearing species, the ecosystem might well continue to function normally. But if one was load-bearing, ecosystem function would change and you might be tempted to conclude that a 20% loss of species alters ecosystem function when, in fact, it was only the loss of one that did so.

Another approach is to look at the correlation between biodiversity and ecosystem function. For example, when a cultivated field is abandoned and allowed to go through ecological succession, net primary productivity initially increases during the early part of succession as both numbers of species and biomass increase. But a point may be reached where productivity (a measure of ecosystem function) levels off even though biodiversity continues to increase, or vice versa. At this point biodiversity and productivity appear to become "decoupled."

Experimental studies in manipulated terrestrial and aquatic ecosystems have supported the hypothesis that diversity is related to such ecosystem parameters as stability, productivity, and nutrient cycling.[19] These studies can easily be summarized as showing that a cluster of species (i.e., a "critical mass") seems essential to ecosystem functioning, as measured by such variables as plant biomass, a reflection of net productivity. In other words, there is a positive correlation between species richness and ecosystem function, at least up to a point.

Ecologists recognize two mechanisms that could account for the correlation between species richness and ecosystem function. The first, called niche determination and facilitation, hypothesizes that species complement one another and the combination of species is thus able to enhance ecosystem function. The second mechanism is stochastic; the argument is that regional and local stochastic processes add species, some of which, by chance dispersal, are highly productive, thus their addition disproportionately enhances ecosystem function. Researchers are quick to note that these two views are opposite ends of a continuum. They are not mutually exclusive. And they are vague.

Researchers also agree that the clearest demonstration of the connection between biodiversity and ecosystem functioning would be "overyielding," a term that means that the productivity of a mixture of species would always exceed that of any one of the species when grown alone.

Overyielding was tested in caddisfly species.[20] Caddisflies are insects (Trichoptera, Hydropsychidae) with aquatic larvae. The larvae are suspension feeders, building a silklike net in a streambed that acts to filter particulate matter that is then devoured by the larvae. Three species of caddisfly, *Hydropsyche depravata*, *Cerato-*

psyche bronta, and *Cheumatopsyche sp.*, each build a different kind of feeding net. Researchers tested each species alone and in combination with the other two. The single-species tests consisted of 18 larvae of a single species and the combination tests consisted of 6 larvae of each species. The variable measured was the amount of filtered suspended particulate matter (SPM). The result was that streams with a mixed assemblage of caddisfly larvae had a 66% greater consumption of SPM compared with the total consumption in all species monocultures. The reason why the mixed assemblages outperformed any of the monocultures was related to the effect on stream flow. The species richness had the effect of decreasing the deceleration of water current. Faster water current provided more SPM per unit time, thus enhanced the amount captured by the mixed assemblage. Each species of caddisfly inadvertently facilitated the delivery of resources to the other two species due to biophysical interactions relating to enhancing stream flow rate. None of the monocultures equaled the mixed assemblage, so, in this experiment, there was clearly overyielding due to biodiversity. But such overyielding says nothing about stability or resistance to perturbation.

Overyielding was also demonstrated in a seven-year study of grassland plots at Cedar Creek, Minnesota.[21] Researchers established 168 plots, each 9 by 9 meters. They seeded the plots in May of 1994, so that each plot had either 1, 2, 4, 8, or 16 species, each with replicates ranging from 29 to 39 depending on species combination. The actual species mixture per plot was random, from a pool consisting of 18 grassland perennial plant species that included eight grasses, four legumes, four non-legume forbs, and two woody species. Each of the species was tested at least once in monoculture. Aboveground biomass and total (including roots) biomass were measured for each plot over the years of the study.

In 1999 and 2000 the number of species in the plot exerted strong positive effects on both aboveground and total biomass. For example, in 2000, the sixteen-species plots had 22% greater aboveground biomass than the eight-species plots. The sixteen-species plots averaged 39% greater aboveground biomass and 42% greater total biomass than the very best of the monoculture plots.

Even the legumes (which fix nitrogen), when grown as monocultures, failed to equal the mixed species plots.

Productivity is but one parameter of ecosystem function. Stability is another. Robert May's theoretical analysis of the relationship between biodiversity and ecosystem function concluded that there is no firm relationship between biodiversity and ecosystem stability.[22] In other words, more species, as such, do not necessarily result in an increasingly stable ecosystem, one that resists perturbations. But that was a theoretical analysis. What do the data show?

One of the difficulties with the stability concept is that it has had various meanings in the ecological literature.[23] For several decades, as I did with my Ph.D. work, ecologists have engaged in a "diversity-stability debate" in an attempt to ascertain exactly what the relationship is between biodiversity and ecosystem stability. It is critical to be precise in what question is being asked with relation to testing the impact of biodiversity on ecosystem stability.[24] Early successional ecosystems may rebound from perturbations more rapidly than older ecosystems but be less resistant to invading species than older ecosystems. Thus early successional ecosystems by one measure are more stable (they rebound more quickly) but by another measure are less stable (they are more apt to be invaded).

There is theoretical and some experimental support for the "diversity resistance hypothesis," asserting that more diverse communities resist invasion more successfully than less diverse ecosystems. In another study at Cedar Creek, Minnesota, the diversity resistance hypothesis was tested by a team of researchers.[25] The researchers studied 147 experimental grassland plots and followed the establishment and success of 13 species of exotic non-native plants ranging from grasses to legumes. They employed a method termed "neighborhood analysis," which looks at the relationship among three variables: number of plant species within the neighborhood, total number of plants within the neighborhood, and an index of crowding that considers both distance and size of all plants within the neighborhood. The experimental design comprised 147 different plots, each 3 by 3 meters. Each plot began with either 1, 2, 4, 6, 8, 12, or 24 species drawn randomly from a pool of 24 species. Researchers recorded the spatial coordinates and size of approximately 53,000 individual plants, 40,000 native resident species, and 13,000 invader species (mostly Eurasian) over a two-year period. Prior to the beginning of the study, the plot had been weeded and disced to remove seeds of invading species so that when invaders arrived, it was from seed dispersal.

The results supported the contention that species richness protects against invasion by alien species. Plot species richness significantly reduced the amount of cover by invader species as well as maximum invader plant size. There were 91% (in 1997) and 96% (in 1998) reductions in invader cover in plots where there were 24 species compared with plots that were monocultures. Increasingly dense, species-rich, and crowded neighborhoods were strongly influential in preventing invading species from attaining large size. The researchers concluded that "local biodiversity represents an important line of defense against the spread of invaders."

The actual mechanisms underlying the effect of biodiversity on ecosystem function are complex, and many involve interactions between trophic levels (top-down and bottom-up) as well as within trophic level such as competition among species. A study by a team of researchers demonstrated that soil invertebrate fauna enhanced grassland succession and diversity.[26] If plant diversity is related to ecosystem function, as shown by examples cited above, but plant diversity is affected by the influence of soil fauna, then soil fauna is ultimately responsible for ecosystem function.

Food web structure can affect plant diversity in numerous ways. For example, herbivores may feed mostly on dominant plant species (because insect herbivores tend toward specializing on specific plant species or families, thus are attracted when their food plants are abundant), thus providing subordinate species with some relief from competition. Symbiotic fungi below ground may selectively benefit certain plant species by aiding in the root systems' uptake of nutrients. By the same token, root pathogens may exert strong effects on dominant species, helping to maintain biodiversity.

In the study just described, soil fauna was added to experimental grassland ecosystems established in sterilized soil. The soil fauna consisted of nematode worms, micro-arthropods (such as springtails and mites), and beetle larvae. The grassland plots represented early, mid, and late successional plant species assemblages. When soil fauna was added to each of the three successional grassland communities, dominance patterns among plants shifted. Specifically, plant species typical of late succession attained greater dominance. In the control plots, devoid of the added soil fauna, plant species more typical of mid-succession dominated. The mechanism underlying the pattern was that root biomass was reduced among early and mid-successional plants but it increased among late successional plants. In other words, the soil fauna was selectively feeding on roots of early and mid-successional species, thus allowing the late successional species to grow more rapidly. In addition, the presence of the soil fauna enhanced plant species diversity as the proportion of dominant plant species decreased. It was clear from this study that soil fauna directly affected plant species richness as well as increasing the rate of succession, all because of selective root herbivory on dominant plant species.

One of the most compelling papers about the potential relationship between biodiversity and ecosystem function is a 2006 study by Boris Worm et al. titled "Impacts of biodiversity loss on ocean ecosystem services."[27] This remarkable study contains the pessimistic warning that virtually all commercial fish and other seafood species will be depleted by mid-century. But the researchers also assert, with hard data taken from 32 small-scale experiments as well as other, larger scale studies, that restoring biodiversity to areas of the oceans where it has been impoverished will allow the full recovery of productivity and stability. The paper shows a compelling correlation between biodiversity and ecosystem function but falls short of identifying mechanisms to fully explain the relationship. It does suggest that global concerns about fisheries stability should focus intently on biodiversity maintenance.

In a review paper discussing what he terms "newly emerging paradigms" in ecology, Shahid Naeem offers the argument that biodiversity does, indeed, govern ecosystem function.[28] Naeem regrets that the "tone" of the debate among professional ecologists over the influence of biodiversity on ecosystem function has been so strident. It represented a philosophical dialectic tension between those who "seek to explain nature by studying its parts and those who seek to explain nature by studying whole-system processes." The outcome of such debates is nonetheless a productive flurry of research activity, as has been the case with biodiversity and ecosystem function. Naeem demonstrates how the paradigm of nature, as he defines it, has evolved since the time of the ancient Greek scholars to become what is known today as BEFP, the "biodiversity-ecosystem function paradigm," or, more succinctly, BEF. To me, all BEF says is that many species are better than some species and some species are better than just a few species. But maybe that says a lot, and maybe ecologists need to get that word out better than they have. Research on BEF is now focused on determining how biodiversity affects the performance level, stability, and redundancy of multiple ecosystem functions, rather than attempting to tease out the effects of diversity on single functions such as primary productivity or biogeochemical cycling. The results show that overall ecosystem functioning is more sensitive to species loss than single functions, and that biodiversity really represents "multifunctional complementarity" among species.[29] In other words, biodiversity is highly interactive with ecosystem functioning, different species having different impacts, but all species having some impact. This approach is promising, but the authors urge that more research is needed, especially on natural ecosystems rather than experimental assemblages.

So has ecology found its paradigm in BEF? Has it abandoned the balance of nature for the importance of biodiversity as the fiber that maintains ecosystem functions? Perhaps. As my mother used to counsel me, "time will tell." Hopefully that time will be sooner than later.

~ 14

Facing Marley's Ghost

Remember that old bumper sticker, "Have you thanked a green plant today?" Well, to bring it into this century, what have Earth's ecosystems done for you lately? Drawing a blank? You're not alone. Most people regrettably have only the foggiest notion. But consider how each of us benefits from all those oceans, estuaries, marshes, fields, forests, savannas, grasslands, deserts, etcetera. Taken together, they provide "nature's services," the natural ecosystem functions upon which all life, including humanity, ultimately depends. Ranging from purification of air and water, cycling and movement of nutrients, climate modification, generation and preservation of soils and renewal of soil fertility, to seed dispersal, pollination of crops and other vegetation, to maintenance of biodiversity (including the esthetic satisfaction it provides), it is clear that the functioning of natural ecosystems is essential to human welfare.[1]

I am coming up on four decades of teaching ecology to people ranging in age from 18 to about 25. They are smart, eager college students and most are not cynical. Given the subjects I teach, as well as my relatively advanced age, my students frequently ask my opinion about the future, their future. Most of my current students, with health and luck, may expect to approach the turn of the next century. After devoting my professional career to the study of ecology and evolutionary biology, I am hesitant to provide these hopeful young people with unbridled optimism. President John Kennedy once said, in essence, that global problems are created by humans and thus humans should be able to solve them. (The actual quote is "Our problems are man-made, therefore they may be solved by man.") Sounds good, and I often use that thought to encourage my students to think with hope and optimism. What does bother me about Kennedy's words is that I confess to the belief that global problems are far easier to create than to solve. At least that seems to have been the case for the past 10,000 years.

I was in graduate school in 1970, the year of the first Earth Day (April 22). Public interest in the environment was suddenly high, and my professors, as well as ecologists throughout this nation, were sought out by the media for sound bites and, on occasion, extended interviews. What about pollution, biodiversity, pesticides, extinction, human population growth? How deeply was the world in trouble? Would ecology overtake economics in the new world order?

The holistic nature of ecology does, indeed, lead to a different view from that traditionally associated with Western economic pursuits. For this reason, ecology was dubbed by ecologist Paul Sears as the "subversive science."[2] Sears asked, "Is ecology a phase of science of limited interest and utility? Or, if taken seriously as an instrument for the long-run welfare of mankind, would it endanger the assumptions and practices accepted by modern societies, whatever their doctrinal commitments?" In other words, is ecology a potential destroyer of societal paradigms?

When the first Earth Day provided them with a forum, ecologists had little to say, never mind of a subversive nature. Oh, mind you, they said plenty, but when those words were carefully parsed, most predictions and dire warnings, to say nothing of pontifications, were based on sparse data, nascent theory, and a whole lot of exaggerated rhetoric. This was not the fault of the ecologists, who were sincere in responding to public interest in what they had to say, even as they were overreaching in saying it. They were savoring their fifteen minutes of fame and doing the best they could with what little solid data they had. Today the situation is greatly improved. Ecology has become sufficiently robust over the decades since Earth Day 1 to become a relatively predictive science that can provide prescriptive solutions to at least some daunting environmental problems. But that fact, of course, does not necessarily translate into action. Ecologists still have a long way to go to "get the word out" and even further to be taken seriously. Ecologists as a group remain a minor part of a far broader and strongly anthropocentric global society. It's still "all about us." The philosophical dualism between humanity and the vast assemblage nonhuman life forms of Earth persists, indeed thrives. And it has consequences.

Such an anthropocentric view was once defensible. Nature is tough. It was not easy to convert forest into pasture or a crop field, or build a house, or fish the open seas. Beyond that, the untamed vastness of nature was considered as frontier, which people had a right to exploit for their personal gain. Nature was perceived as immense, its resources inexhaustible, or so it naïvely seemed to most people. The philosopher John Locke, whose views were pivotal in framing the philosophy upon which the United States was founded, argued that humans have a moral responsibility to improve and to "civilize" the landscape and that ownership of the land is valid only when the land is so "improved."[3] Locke viewed nature as without intrinsic value until it was altered, improved, enhanced by human labor. How anthropocentric is that!

The historian Lynn White, Jr., in an influential essay,[4] argued that Judeo-Christian theology promoted the view common to Western societies, that nature is fundamentally distinct from humanity (which was, of course, uniquely created in the image of God). As such, we view ourselves as profoundly "apart" from nature. Of course Darwin thoroughly refuted this view, but virtually all philosophers as well as others outside of science (and many inside of science) paid little or no attention. The strong tradition of the anthropocentric paradigm still prevails throughout the world, not only in Judeo-Christian societies but in essentially all industrialized societies.

Garrett Hardin brilliantly focused upon the commons example as an analogy for Earth's threatened sustainability. A commons is an area owned by no one but available for use by all. In colonial New England, for example, the commons was a place where all villagers could graze their livestock, the assumption being that the productivity of the commons exceeded the grazing pressure from the livestock. Only while such an assumption remained true was the commons viable. In a widely read essay titled "The Tragedy of the Commons,"[5] Hardin equated the Earth's water, atmosphere, soil, and other resources including biodiversity with the commons concept as it was traditionally applied to such activities as cattle grazing and freedom of the seas. He noted that as long as human population or the rate of exploitation is relatively low compared with the capacity of the commons in question, the resource could, indeed, be treated as a commons, open to all, owned by none. But when human population and technology exceeds the carrying capacity of the commons, "tragedy" ensues in that there is no incentive for the various exploiting parties to cease what has, by then, become overexploitation. Indeed the contrary is true. It becomes in the best interest of the most affluent parties to increase rate of exploitation, maximizing their gain, since in the short run they gain more than they lose, even as the commons is more quickly destroyed. Hardin's conclusion was that once the commons is no longer viable as a commons, only coercion succeeds in maintaining the resource. Interested parties must agree to abide by strict rules (laws) with punishment for those who violate them.

The most essential development of the twentieth century was, in my opinion, the loss of global commons (in its many forms) as a utilitarian reality. No longer are any oceans, lakes, forests, grasslands, or deserts exempt of human influence. The same is true of the global atmosphere. The commons concept simply no longer applies anywhere. It has been replaced by the realization that human activity is now causing overall alteration and in many cases degradation of virtually all global ecosystems. Earth's very sustainability is at issue (see below). This is certainly due to economic practices that reflect Adam Smith's capitalistic free-market views (even within socialistic countries, such as China), coupled with lack of insight about the value of the services provided by natural ecosystems, services that are implicit in the commons concept.

How should human impact on global ecosystems be measured? The density of the global human population, now at about 6.7 billion, alone does not measure the cumulative effects of humanity on the Earth's ecosystems. This question is obviously more complex and could, in theory, involve thousands of variables. Human impact is obvious when a woodlot or forest is cut to make room for agriculture or for a housing development. It is less obvious when nitrogen content in streams and rivers (such as the Mississippi) increases because of fertilizer runoff, eventually reaching the sea and creating "dead zones," where oxygen has been depleted as a result of microbial stimulation brought on by the added nitrogen input. It is even less obvious when numerous species of oceanic creatures such as porpoises and albatrosses are killed as "bycatch" (non-target species) in immense drift nets, even as fishing stocks continue to decline. Nonetheless, assessing global human impact is a critically important question, as its answer clearly affects the future of the planet.

It is possible to gain some understanding by considering the formula $I = P \times A \times T$, what has become known as the "IPAT formula."[6] I stands for impact, P for population, A for affluence, and T for technology. Note that the variables are multiplied, not added. What this means is that each variable, as its value rises, has a strong effect on accelerating impact. It is obvious that increasing population will inevitably result in added impact. But population interacts with technology and affluence. Wealthy individuals consume far more (in the form of resources such as oil) than poorer persons. Technological advances permit more widespread alteration of ecosystems, especially in rich nations.

The IPAT formula is obviously an oversimplification, but it does show that population alone is not responsible for the total ecological impact humans exert on their planet. Because of industrialized agriculture, humanity now controls something close to 40% of global net primary productivity.[7] That dominance would be impossible without post–Industrial Revolution technologies. Oil now makes corn.

Though *ecology* and *economics* share the same Greek root, *oikos*, there has been a clear gulf between how ecologists and economists interact to share the insights of their disciplines. Recently attempts have been made to combine ecology and economics in ways that might provide both understanding and guidance for future actions. One such study[8] focused on calculating the "ecological footprint" of humans. The ecological footprint is defined as the total area required to produce the food, goods, and other resources an individual consumes annually. Edward O. Wilson,[9] citing widely known data, stated that the ecological footprint for the United States equals 24 acres (14.7 for energy, 3.6 for crops, 0.9 built area, 0.8 pasture, 0.8 fishing grounds), which represents by far the largest ecological footprint of any country. For Germany, like the United States an industrial, developed country, the EF is 11.6, and for Mozambique, an undeveloped country, it is 1.2. The global average is 5.6 acres. This means that, on average, one American citizen is equal to about 4.3 non-Americans with regard to how much he or she takes from the global environment and the impact he or she has upon it. It would require a family with 40 children in Mozambique to exert as much collective environmental impact as a family with two children in America.

Peter Vitousek et al.[10] summarized the collective effects of human domination of Earth's ecosystems. The study recognized the following general categories: land transformation, impact on oceanic (particularly coastal) ecosystems, alteration of biogeochemical cycles, and biotic changes such as loss of biodiversity and increasing prevalence of invasive species. Each of these categories includes many of Earth's services, which, I repeat, are functions supplied by natural ecosystems that are normally not accounted for in economic analyses (see below). They appear to be "free" as long as the ecosystems remain intact. Now they are being undermined.

Humans lay claim to over 40% of the total net biological productivity on land and are using about 50% of the renewable freshwater. Carbon dioxide and other greenhouse gases continue to rise steeply, with carbon dioxide up by nearly 35% since 1800, and methane 135% above historic natural levels. Human activities have profoundly altered aspects of the global nitrogen cycle, mostly due to increasing use of fossil fuels.[11] There has been ozone depletion in the upper atmosphere, an event clearly tied to the emission of chlorofluorocarbon compounds. Earth's climate is changing, almost certainly forced by anthropogenically released greenhouse gases. There has been an estimated loss of about 20% of the world's topsoil over the past 50 years and about 20% of agricultural land over the same period, largely due to overuse. There continues to be loss of biodiversity from habitat deterioration and invasive species. An evaluation done by Birdlife International[12] notes that of the world's total of 9,797 bird species, 1,186 (12%) are threatened with extinction within the present century. And that's not all.

S. R. Palumbi[13] has described how humans have accelerated the evolutionary process in bacteria and viruses, noxious weeds and arthropod pests, and even altered life histories of fishery species. Taken together, human impact on Earth is staggering and grows daily.

Vitousek et al.[14] offer three suggestions for how to approach dealing with human domination of Earth's ecosystems. First, they suggest that the ecological footprint of humanity be stabilized. This would be accomplished by reduced rates of exploitation. Reduction of greenhouse gas emissions is an obvious example. Second, it is essential to accelerate basic ecological research to gain a far more precise understanding of Earth's ecosystems and how they interact with the numerous components of human-caused global change. For example, it remains unclear as to the relationship between biodiversity and ecosystem functions such as net productivity, and ecosystem stability. Third, humans must accept that the time has been reached when humanity should responsibly manage the planet. There is no natural balance of nature, either in the past or present, and, given how profoundly humans affect Earth's ecosystems, we must now make choices and manage ecosystems not only for the flow of goods and services provided to humanity but also for maintaining populations, species, and diverse ecosystems.

The question for all global citizens thus becomes how to ensure the sustainability of Earth's numerous and essential ecological services without potentially catastrophic disruption. This is the question my students will face throughout their lives. It may seem incredible that such an issue even be posed, but reality suggests otherwise. Conservation now means so much more than the protection of endangered species or the creation of wildlife sanctuaries. Conservation now deals with the entire breadth of the land and seascapes, how to bring economic, social, and environmental interests (often called "stakeholders") together in such a way that ecosystem services will continue to the collective economic benefit of society. The science of conservation biology strongly overlaps with theoretical ecology, population genetics, landscape ecology, restoration ecology, and ecosystem management. The objective of such efforts has increasingly focused on developing strategies that result in sustainable resource use.

In remarks to the American Association for the Advancement of Science, ecologist Peter H. Raven[15] (who was President of AAAS at the time) said, "We must find new ways to provide for a human society that presently has outstripped the limits of global sustainability." Raven noted that only 4.5% of the global human population resides in the United States, even though this country controls 25% of the world's wealth and produces between 25% and 30% of its pollution. Inequities between nations of wealth and those seeking development underlie many actions that result in negative effects on ecosystems, including overexploitation of the global commons.

To reach sustainability, it is obviously necessary to recognize the need for strong cooperation among all interests. Using game theory and simulation, it can be demonstrated that reciprocal cooperation, based merely on establishment of a reputation for honorable reciprocity, will overcome the tendency toward greed that degrades the commons. Indeed the simulation demonstrates that it is possible for the commons to be productive and stable.[16]

But it is a long way from game theory simulation to global politics and economics. Realities are, after all, realities. Given the political realities of today's world, how might nations approach the sustainability of Earth's ecosystems? Perhaps the best argument for discouraging selfishness among nations is ultimately to appeal to that very quality, selfishness. Given that there is but one Earth, it is in every person's (and thus in every government's) self-interest to act in a manner that contributes to sustainability.[17] But for this obviously logical notion to prevail, it will require an expansion of economics to include global sustainability as an objective. It will also require the inclusion of environmental considerations in ethical judgments. These requirements are daunting, as they amount to societal paradigm shifts.

How do we begin to attack such problems? Perhaps by asking just how much, in dollar value, are nature's services worth?

A team of researchers led by Robert Costanza[18] used numerous databases to estimate that nature's services, presuming that humans had to somehow "pay" for them, would be worth about 33 trillion dollars annually. It is noteworthy that this estimate is about double the gross world product, the total of all gross national products, which, at that time, was about $18 trillion. Balmford et al.[19] updated the estimates to the dollar's worth in 2000 and suggested a rough average of $38 trillion, with a range of between $18 and 61 trillion. The wide range reflects the extraordinary difficulty of attempting to measure the macroeconomic worth of all ecosystem goods and services.

Some economists have criticized the effort by Costanza et al., asserting, among other reasons, that the extrapolations made were inconsistent with accepted principles of economic theory. In response, a team of nineteen researchers headed by A. Balmford, including economists and ecologists (Costanza among them), examined over 300 specific case studies in an attempt to compare what they termed "marginal values of goods and services delivered by a biome when relatively intact, and when converted to typical forms of human use."[20] Their results were consistent in showing that natural ecosystems have the potential for greater societal economic gain than do ecosystems converted for narrow economic objectives.

In one example it was clear that reduced-impact logging in Malaysian tropical forests did not provide the immediate economic benefits to individuals that would be obtained by high-intensity unsustainable logging, which is what is normally done there. However, unsustainable logging ultimately reduced social and global benefits through loss of forest products (other than timber), flood protection, carbon sequestration, and endangered species. The total economic value of the forest was about 14% greater when managed to be sustainable, using reduced-impact logging techniques.

In a second example, a mangrove ecosystem in Thailand was converted to aquaculture (shrimp farming), something that is occurring in many places in the world's tropics. There was no question that short-term economic interests were well served by such a conversion. However, when social benefits of leaving the mangrove ecosystem intact were included in the economic analysis, the result changed. Benefits such as the sequestration of carbon, storm protection, and protection for fish added much more value to the intact mangrove forest than to the aquaculture ponds. The estimate showed that the intact mangrove forest was worth about 70% more than the aquaculture. Note that temporal scale effect is a strong factor. Conversion of mangrove forest to agriculture happens in a short time span and profits are generated accordingly. The time span necessary to enjoy full benefit of services provided by intact mangrove forest is much greater, the profit far more spaced out, like a small annuity compared with a windfall profit.

Other analyses showed that intact wetland in Canada was worth considerably more than if drained and converted to intensive farming. Sustainable fishing on a Philippine coral reef far exceeded the economic value of destructive fishing. The conversion of a tropical forest in Cameroon to a plantation actually lost money, while there was economic gain to be had from utilizing the intact forest for either small-scale logging or reduce-impact logging.

The Balmford et al. study also showed that five of six biomes have experienced combined net losses of about 11% over a decade as humans convert natural ecosystems into anthropogenic ecosystems.

The reasons why ecosystem loss continues are not surprising. Ecosystem services remain poorly understood with regard to their actual economic value, and it is often difficult to ascertain economic value for some ecological variables. Further, traditionally economists have regarded ecosystem functions as "externalities," whereas economic benefits from conversion are "internalities." The distinction between externality and internality goes back to the concept of the commons. Internality is profit gained by the individuals who invested in the project. But externalities are costs, borne by all, of habitat damage and loss of ecosystem services that resulted from the conversion. In other words, the investors are not charged, taxed, or otherwise responsible for the deterioration they cause in ecosystem services. Such costs are external to the profits. But enacting laws such as the Clean Air and Clean Water acts changes externalities to internalities. The externalities paradigm has dominated economics for centuries, ever since Adam Smith fundamentally defined capitalism. Finally, ecosystems are often exploited due to inflated claims for short-term profits. As mentioned above, plantations in Cameroon were economic failures. It would have been far wiser to utilize intact forest for reduced-impact logging. The current pushes to drill for oil on the continental shelf and to exploit the Arctic National Wildlife Refuge for oil extraction may be other examples.

Balmford et al. conclude with an appeal for what amounts to a major paradigm shift in macroeconomics, if not in the Western moral view of nature: "Our relentless conversion and degradation of remaining natural habitats is eroding overall human welfare for short-term private gain. In these circumstances, retaining as much as possible of what remains of wild nature through a judicious combination of sustainable use, conservation, and, where necessary, compensation for resulting opportunity costs makes overwhelming economic as well as moral sense." Such an appeal brings us to the subject of environmental ethics.

Ethics is a branch of moral philosophy that deals with what one "ought" to do in various situations, how one "ought" to behave as a member of society, striving to act for the common weal. Ethics is the branch of philosophy that focuses on personal responsibility to one's self and to one's society. Ethics is prescriptive behavior based on what is deemed morally appropriate and honorable. There are certain ancient ethical truths that seem obvious and universal. Ethics includes the concept of rights. One should not commit murder, for example, because persons have the intrinsic right to live.

Ethics changes, it evolves, as societies become increasingly enlightened. It has always been deemed sinful to murder but not always to own another human being. At one time it was not a breach of ethics to consider people of certain races intellectually inferior to those of other races. American Indians, more appropriately termed Native Americans, were not accorded citizenship in the United States until 1924. It was not until August 26, 1920, when the Nineteenth Amendment was ratified, that women were permitted to vote in the United States. It was not until the Civil Rights Act of 1964 that the United States government specifically prohibited discrimination on the basis of race and sex in hiring, promoting, and firing. At one time it was considered ethical to permit small children to work long hours in industrial factories. Ethics expands with enlightenment.

Environmental ethics, in any prescriptive sense, was unknown for most of human history. Some religions practice animism, honoring and affording sacred status to such objects as rocks, water, and various plants and animals. But religious respect aside, it was not until well into the twentieth century that humans became fully alerted to environmental degradation such that action is required. Such need for action ultimately affected ethics. In 1949 Aldo Leopold wrote of the land ethic, a clarion call to recognize the value of ecosystem services and to act in a manner that assures their continuance.[21]

Before Leopold, John Muir (1912) wrote of the grandeur of the Yosemite Valley and the Sierra Nevada mountain range.[22] In the final chapter of Muir's book *The Yosemite*, he made an appeal. Plans were in the works to dam the Hetch Hetchy Valley, then much like Yosemite Valley, a splendid and pristine natural area. Muir wrote,

> Sad to say, this most precious and sublime feature of the Yosemite National Park, one of the greatest of all our natural resources for the uplifting joy and peace and health of the people, is in danger of being dammed and made into a reservoir to help supply San Francisco with water and light, thus flooding it from wall to wall and burying its gardens and groves one or two hundred feet deep. This grossly destructive commercial scheme has long been planned and urged (though water as pure and abundant can be got from sources outside of the people's park, in a dozen different places), because of the comparative cheapness of the dam and of the territory which it is sought to divert from the great uses to which it was dedicated in the Act of 1890 establishing the Yosemite National Park.

Muir's pleas did not turn the tide. The O'Shaughnessy Dam, damming the Hetch Hetchy Valley, was completed in 1923. That is not surprising. In Muir's time the environment still seemed an "inexhaustible" resource and ecosystem services went largely un-recognized. Environmental beauty, esthetics, was considered a distant second priority to industrial development. Times may have changed.

Inspired by Rachel Carson and her book *Silent Spring* (1962),[23] some philosophers have focused on how humans *ought* to interact with and respect the environment.[24] The philosophical obstacle was anthropocentrism, the notion that "only humans are the locus of intrinsic value, and the value of all other objects derives from their contribution to human values."[25] Anthropocentrism allows, for example, for animal protection laws that apply to species such as domestic cats and dogs because they, indeed, have intrinsic value to humans, most specifically to their owners. But an anthropocentric view does not as readily accommodate the value of polar bears, old growth forests, or clean air, though in each case there is no obvious reason why it shouldn't.

Beginning in the 1960s, laws were enacted in the United States to protect endangered species and ensure clean air and clean water. These acts in essence created economic internalities by affording environmental components (air, water, organisms) a form of quasi-rights. Societies inevitably reflect their ethical values in their laws, thus the Clean Water and Clean Air acts, as well as the Endangered Species Act, each reflect changes in how citizens of the United States value their environment. Today most development in the United States, from building houses to the construction of shopping malls and highways, normally requires extensive environmental impact studies before construction can proceed.

Unique among the various ethical arguments is one suggested by Edward O. Wilson, termed *biophilia*.[26] Wilson asserts that human beings enjoy an emotionally based affiliation with nature that is at least to some degree innate. He argues that people typically seek homes in which they have a view of water or scattered trees, an environment that evokes visions of humanity's evolutionary association with open savanna ecosystems. Wilson's view of biophilia could be interpreted as neo-Darwinian ethics, as it asserts an evolutionary basis. If biophilia is real, we humans evolved with a predisposition to value nature, at least esthetically.

It is a fact of evolution that humanity has a genetic kinship through time with all other species ever to have occupied Earth. Biodiversity has always been an obvious part of all hominid evolution. As our intelligence evolved, we came to view nature as familiar and as part of our essence as persons. Thus we should have some vague but nonetheless real emotional connection with the natural world. Such a connection could be viewed, in the light of human intelligence, as prerequisite to bestowing ethical rights (of a sort) upon organisms or even ecosystems.

To put the question bluntly, does humanity have a right to act in such a way as to cause the extinction of, say, polar bears? Today most enlightened people would say no, but polar bears are being threatened by melting of Arctic ice due to global warming. So what do we owe them? What action should we take?

Environmental ethics involves choices, not necessarily absolutes, and those choices must be informed by what is known about how ecosystems and their component organisms function. It is incorrect to argue that we have an ethical obligation to, say, protect Brazilian rainforest in order to preserve the balance of nature. Ecologists now know there is no balance of nature. More rigorous information is needed to make a sound argument for affording protective status on an entire ecosystem. It could be argued, for example, that certain areas within Brazilian rainforest have unique assemblages of endemic species and are thus deserving of protection. This very argument, terming such areas "hotspots" of biodiversity, has been made.

Daniel Janzen[27] wrote that natural ecosystems and their component species should be equated with such societal needs as "libraries, universities, museums, symphony halls, and newspapers." Such appeals to an ethical basis for stewardship of nature may seem rather far-fetched, overly idealistic, and, at worst, highly impractical. But do keep in mind that arguments against slavery and for equal rights, for the right of women to vote, for child labor laws, were once viewed as overly idealistic, far-fetched, and highly impractical. Some would say that arguments against universal health care, gay rights, and environmentalism still are. But to many of us they are not. Times do change and societies do evolve.

Environmental ethics must embrace emotive connections between humans and the natural world as well as recognize that humans are dependent for their welfare on the functions supplied by the natural world. There is thus both an emotional and a pragmatic side to environmental ethics. Humanity ought to act in a manner that sustains its planet. It is in our enlightened self-interest to do so. That said, it will likely be an uphill effort.

Perceiving nature as deserving of ethical consideration is ultimately pragmatic for humanity. It satisfies two major needs common to all human beings. First, it assures that ecosystems are valued for the collective services they provide in maintaining a comfortable, livable environment. Nature thus "pays us back" for affording it protection. Mind you, it won't say "thank you," but it pays us back nonetheless. Second, nature, as Wilson suggests, fulfils a real emotional need among people, one that was less obvious when natural ecosystems were much more extensive than today. Certainly the growth of ecotourism as an industry (as yet largely confined to the affluent in society) in the latter part of the twentieth century would suggest that people feel a need to "see nature" at its best.

When Ebenezer Scrooge faced the ghost of his former business partner, Jacob Marley, Marley's ghost wore a cumbersome chain.[28] As Marley explained to Scrooge, "I wear the chain I forged in life. I made it link by link, and yard by yard; I girded it on of my own free will, and of my own free will I wore it."

Scrooge, in a mere evening, and with the help of Marley's and three other ghosts, got the message. He changed his behavior, his attitude, his life, his very humanity. Today we humans, it may be suggested, continue forging a very formidable chain, and of our own collective free will. Like old Ebenezer we need to change, to visualize our future as Scrooge was able to see his, to evaluate our past as Scrooge did his. We need a visit by Marley's ghost. That, of course, will not happen. For most humans environmental issues tend to be viewed as rather arcane, far less important than other more daily concerns. But a close look at what is happening today shows clearly that environmental issues are becoming daily concerns. Hopefully that will be noticed.

Immanuel Kant wrote in *Critique of Pure Reason* that "ideals conflict with reality because that's how they work . . . they guide and goad us to make reality live up to ideals."[29] That's good metaphysics and bad science. Let me suggest that now is the time in Earth's history when metaphysics needs to yield to science and reason in driving policy decisions. The twenty-first century should be the century of environmental ethics based on scientifically acquired knowledge, the new paradigm. The time is overdue to face ecological reality for what it is.

The choice is ours. Earth will not hate us if we do nothing nor will it thank us if we take action. We will most certainly not kill our planet, but our own future as a species remains insecure. The balance of nature, whatever it is and whatever it will become, is our choice, and, I would argue, our moral responsibility. Our future welfare depends on our actions toward Earth's ecosystems.

Marley's ghost is staring us in the face.

∾ Epilogue

Has ecology found its paradigm, a paradigm that will help take humanity through the twenty-first century? Is biodiversity linked so tightly with Earth's ecosystem-level services that active intervention to maintain biodiversity and thus ensure sustainability will guide human actions in the years to come? Well, maybe. That remains to be seen.

I don't think that what ecologists describe as a paradigm shift really amounts to such a thing. I once did and even titled a talk and paper to reflect that view.[1] But now I think ecologists are just seeing biodiversity in a new light, free from the once looming philosophical assumption of balance. In that sense, ecology's new paradigm is nothing more than an intellectual Neckar cube: you know, when you look at it one way it seems imbedded within the page, but looked at slightly differently, it pops out at you. It's a matter of perception. Ecologists now perceive the workings of nature as free from constraints (and thus false assumptions) imposed by assuming some optimal state of natural balance. Equilibrium models have given ground to stochastic realities. Ecology's paradigm shift is, in my opinion, nothing more than a change of perception, brought about by more robust modeling, lots of data, and a good reality check. That does not mean such a view is unimportant. Just the opposite is the case. Whether ecology has found its paradigm or not, in the broader sense, humanity should adopt a different paradigm in its view of nature and nature's services. That is really the important paradigm shift.

Recall John Terborgh's study of ecological meltdown in fragmented tropical forest (chapter 12). Terborgh makes the point that
once the meltdown is complete, an ecosystem will still be present.
It is no longer what is had been, a diverse ecosystem. It is now depauperate, but it's there. Will we continue to do that to Earth?

I have a colleague in the social sciences who says that people
will always have to come first. To the degree that natural ecosystems can be preserved without undue hardship to people, he agrees
with conservation goals. Such a view represents, yet again, the demonstrably false but persistent philosophical dualism between
humans and nature. Consider, for example, the thousands of
human lives lost due to the cyclone that struck Burma in May of
2008, lives that might not have been lost if mangrove forest had
been preserved as a buffer against such occasional meteorological
occurrences. But in Burma, on the Irrawaddy Delta, mangrove
forest had been eliminated, replaced by shrimp farming and rice
paddies. Still, my colleague likely represents the majority view
among Earth's current human inhabitants. As I write this during a
year when a presidential election will be held in the United States,
there is precious little mention of environmental concerns, including global climate change, by the presidential candidates. They
appear not to have ever heard of ecological footprints, though a
former vice president (who was once a presidential candidate) will
hopefully get their attention on such matters. At least Al Gore is
trying.

My students, their generation, will have to solve an ethical dilemma. They will have an ever more informed and holistic view of
the function of diverse nature and how it provides and maintains
services essential to human welfare. They will grapple with the rights
of creatures that are not human, rights to mere existence. They will
need to balance those insights with the demands of a global human
population that will likely reach 8–10 billion by mid-century. If
they succeed, a really meaningful balance of nature, one quite distinct from the original concept, may be achieved. Their choices,
their actions, though informed by science and by ecology in partic-

ular, are nonetheless largely ethical societal decisions. Ought ethics expand to encompass ecosystem-level concerns? If so, will it?

My advice to my students echoes what my freshman history professor, Mr. Syme, signed in my yearbook.

"Do right."

❧ Acknowledgments

My career in ecology now spans over four decades, and many individuals have influenced me. My debt to them, particularly those who mentored me, is immense, and I will not begin to try and name them all. But for many years I have had long and productive talks with Don Shure, Frank Kuserk, and Dean Cocking at annual meetings of the Ecological Society of America, and some of what we discussed is undoubtedly contained herein. I recall in particular a late night conversation with Frank about just what, if any, paradigms exist in ecology.

Some material in this book was part of a paper I authored that was published in *Northeastern Naturalist*, vol. 5, no. 2 (1998). I am grateful to Joerg-Henner Lotze for permission to incorporate some of that material into this book.

I also thank John Alexander of Recorded Books for permission to adapt some material from two of my published Modern Scholar lecture series, "Behold the Mighty Dinosaur" and "The Ecological Planet: An Introduction to Earth's Major Ecosystems," for use in this book.

I wrote this book while having the honor of being A. Howard Meneely Professor of Biology at Wheaton College, which facilitated a full-year sabbatical. I thank Provost Susanne Woods and Provost Molly Easo Smith of Wheaton College for providing me with the time necessary to write. The following persons have kindly read and commented on various chapters: Martha Vaughan, William E. Davis, Jr., Betsey Dyer, Donald Shure, Taber Allison,

Geoffrey Collins, Scott Shumway, and Robert Askins. Their comments and suggestions were much appreciated, but any errors or other difficulties with this book are entirely my responsibility. Above all others, I deeply appreciate the support of my editor Robert Kirk of Princeton University Press. His enthusiasm and encouragement turned a long-simmering idea into a book. Thanks again, Robert.

Notes

CHAPTER 1 WHY IT MATTERS

1. The quote that begins the chapter is from Stuart L. Pimm's book *The Balance of Nature? Ecological Issues in the Conservation of Species and Communities* (Chicago: University of Chicago Press (1991), p. 4.

2. Several good books each present a fine overview of the various aspects of astronomy and cosmology discussed in this chapter. For example, see F. Adams and G. Laughlin, *The Five Ages of the Universe* (New York: The Free Press, 1999); B. Green, *The Elegant Universe* (New York: W.W. Norton, 1999); T. Ferris, *The Whole Shebang* (New York: Simon and Schuster, 1997); M. Rees, *Just Six Numbers* (New York: Basic Books, 1999); and L. Smolin, *The Life of the Cosmos* (New York: Oxford, 1997).

3. See http://map.gsfc.nasa.gov/universe/uni_age.html for a discussion of how the age of the universe is calculated.

4. Images of the collision are available at http://nssdc.gsfc.nasa.gov/planetary/sl9/comet_images.html.

5. There is abundant documentation of this reality of mummy preparation. See, for example, http://www.civilization.ca/civil/egypt/egcr06e.html.

6. There is an abundance of websites and books dealing with the naturalistic fallacy. See, for example, http://www.iscid.org/encyclopedia/Naturalistic_Fallacy.

CHAPTER 2 OF WHAT PURPOSE ARE MOSQUITOES?

1. See Richard Dawkins's brilliant book *The Selfish Gene* (Oxford: Oxford University Press, 1989).

2. Hominins include species such as *Australopithecus afarensis*, etc. We are *Homo sapiens*, the one and only "wise man." It is not doubtful that other hominins were "smart," but we are likely intellectually well beyond them.

3. See http://www.nature.com/nature/journal/v400/n6747/full/400861a0.html for a paper that tracks the evolution of resistance to an insecticide in a mosquito species.

4. See http://www.fond4beetles.com/prologue.html for more on Haldane and his inordinate fondness for beetles.

5. The film is available on DVD, rated 4.8 out of 10 at http://www.imdb.com/title/tt0081353/.

6. Charles Darwin wrote an entire book on this subject, *The Various Contrivances by Which Orchids Are Fertilized by Insects.*

7. See S. A. Levins, "The Problem of Pattern and Scale in Ecology," *Ecology* 73, no. 6 (1992): 1943–67.

Chapter 3 Creating Paradigms

1. See F. N. Egerton, "Changing Concepts of the Balance of Nature," *Quarterly Review of Biology* 48 (1973): 322–50, for a thoughtful and detailed review.

2. The most well-known reference for scientific revolutions is Thomas S. Kuhn's *The Structure of Scientific Revolutions.* There are numerous internet sites devoted to Kuhn's work, and the book is readily available. See also I. Bernard Cohen, *Revolution in Science* (Cambridge, MA: Belknap Press, 1985).

3. R. Feynman, *The Pleasure of Finding Things Out* (New York: Perseus, 1999).

4. See http://plato.stanford.edu/entries/popper/ for more on Popper.

5. Many internet sites discuss this quote. See, for example, http://www.eequalsmcsquared.auckland.ac.nz/sites/emc2/tl/philosophy/dice.cfm.

6. See, for example, the book *Summer for the Gods: The Scopes Trial and America's Continuing Debate over Science and Religion* by Edward J. Larson (New York: Basic Books, 2006).

7. See Sagan's wonderful book and television series *Cosmos*, first aired in 1980. The series is available on DVD.

8. Darwin argued in *The Descent of Man* (1871) that such uniquely human traits as morality had their evolutionary roots in the social instincts, which he compared among various social animal species. It was just over a century before the subject was again seriously treated in Edward O. Wilson's sweeping book *Sociobiology: The New Synthesis* (Cambridge, MA: Belknap Press, 1975).

9. The film is available on DVD. See http://www.filmsite.org/unfo.html for a detailed description of the film, including the line quoted in the chapter.

10. I recommend David C. Lindberg's fine book, *The Beginnings of Western Science* (Chicago: University of Chicago Press, 1992) for a more detailed treatment of early Greek science.

11. See Egerton, "Changing Concepts of the Balance of Nature" (n. 1 above), for a wonderful and detailed treatment of this and other examples.

12. See Lindberg, *The Beginnings of Western Science* (n. 10 above).

13. See Mayr's book *The Growth of Biological Thought* (Cambridge, MA: Belknap Press, 1982). The quote is from p. 304.

14. See Egerton, "Changing Concepts of the Balance of Nature" (n. 1 above).

15. The classic reference on the *scala naturae* is the book by A. O. Lovejoy, *The Great Chain of Being* (Cambridge, MA: Harvard University Press, 1936).

16. Mayr, *The Growth of Biological Thought*.

17. Emphasis mine. See Lindberg, *The Beginnings of Western Science*, p. 54.

18. See Mayr, *The Growth of Biological Thought*, p. 37.

CHAPTER 4 ECOLOGY B.C. ("BEFORE CHARLES")

1. Ernst Haeckel (1834–1919) was an influential and controversial figure in biology. Though he made numerous contributions, including describing and naming many species, he is best known for his assertion that "Ontogeny recapitulates phylogeny," derived from his detailed studies of embryos and their development. Haeckel's views on the evolution of race have been cited as inspiring Hitler's naïve and tragic policies. Blame Hitler, not Haeckel.

2. Page 62 of the first edition of *On the Origin of Species*.

3. See P. Ball, "Triumph of the Medieval Mind," *Nature* 452 (2008): 816–18.

4. Two books document the rise of science as well as the emergence of various scientific revolutions. See D. C. Lindberg, *The Beginnings of Western Science* (Chicago: University of Chicago Press, 1992), and J. B. Cohen, *Revolution in Science* (Cambridge, MA: Belknap Press, 1985).

5. See the classic book by John C. Greene, *The Death of Adam* (Ames: Iowa State University Press, 1959), for an engaging chapter on how Europeans reacted to apes.

6. See Richard Conniff's article, "That Great Beast of a Town," *Natural History* 117, no. 2 (March 2008): 44–49, for a full discussion of the impact that newly discovered animals had on the London citizenry.

7. *The Autobiography of Charles Darwin* (New York: W.W. Norton, 1958), p. 72.

8. There are numerous websites devoted to Leeuwenhoek. See, for example, http://www.ucmp.berkeley.edu/history/leeuwenhoek.html.

9. See http://www.amnh.org/education/resources/rfl/web/essaybooks/earth/p_hutton.html.

10. Cuvier (1769–1832) invented the science of comparative anatomy and proved beyond doubt that extinction really happened, and happened rather routinely. See http://www.ucmp.berkeley.edu/history/cuvier.html for more.

11. Jefferson confronted the question of extinction when he examined bones from an extinct ground sloth, *Megalonyx*, that he believed to be a kind of lion. See http://gsa.confex.com/gsa/2005AM/finalprogram/abstract_97116.htm for more.

12. John Ray (1628–1705) is discussed on numerous websites. For a good overview, see http://www.ucmp.berkeley.edu/history/ray.html.

13. For a good biography of Linnaeus, see W. Blunt, *Linnaeus: The Compleat Naturalist* (Princeton, NJ: Princeton University Press, 2002). Also see http://www.ucmp.berkeley.edu/history/linnaeus.html.

14. See http://www.ucmp.berkeley.edu/history/paley.html for a brief biography of Paley.

15. The most recent biography of Lamarck is by A. S. Packard, *Larmarck, the Founder of Evolution: His Life and Work* (Gloucester, UK: Dodo Press, 2007). Lamarck is the subject of many websites, but see http://www.ucmp.berkeley.edu/history/lamarck.html.

16. Darwin published the *Origin* in 1859 and devoted the remainder of his life to publishing books on various aspects of evolution by natural selection. The process of heredity troubled him deeply, particularly what he believed to be inheritance of what he called "use and disuse." He thought there must be some mechanism by which heredity responds to use and disuse, hence pangenesis was invented. He first described it in his book *The Variation of Plants and Animals under Domestication*, published in 1868.

17. August Weismann (1834–1914) demonstrated clearly that Lamarck's view of genetics was flat-out wrong. But the idea of inheritance of acquired characteristics was appealing and persisted. In the twentieth century, Trofin Lysenko directed genetics research in the Soviet Union under Stalin, research based on the assumption that Lamarckism actually worked. Russian genetic research was, as a result, set back for decades.

18. The first to suggest the Red Queen arms race metaphor was Leigh Van Valen. See Van Valen, "A New Evolutionary Law," *Evolutionary Theory* 1 (1973): 1–30. But before Van Valen, Ronald Fisher, one of the key figures in the neo-Darwinian synthesis of the 1930s, described a similar aspect of natural selection in his classic book, *The Genetical Theory of Natural Selection* (Oxford: Oxford University Press, 1930).

19. Go to http://www.todayinsci.com/W/White_Gilbert/White_Gilbert.htm for an overview of White's life plus a link to the text of *The Natural History of Selborne*.

CHAPTER 5 ECOLOGY A.D. ("AFTER DARWIN")

1. There are numerous biographies of Charles Darwin. I regard two as definitive. The first is the two-volume work by Janet Browne. Volume 1 is *Charles Darwin: Voyaging* (1995), and volume 2 is *Charles Darwin: The Power of Place* (2002). Both were published in hardcover by Alfred A. Knopf and both are available in softcover editions from Princeton University Press. The other recommended biography is by Adrian Desmond and James Moore, *Darwin: The Life of a Tormented Evolutionist* (New York: Warner Books, 1991). For a briefer but nonetheless insightful treatment I strongly recommend *Darwin, Discovering the Tree of Life*, by Niles Eldredge (New York: W.W. Norton, 2005). Also see http://www.aboutdarwin.com/; it is extremely complete and has numerous links to other sites including the text of some of Darwin's books.

2. Several thorough biographies of Wallace have been published recently. I recommend *Alfred Russel Wallace: A Life* by Peter Raby (Princeton, NJ: Princeton University Press, 2001). Also see http://www.wku.edu/~smithch/index1.htm.

3. It should be noted that Wallace seemed to exhibit nothing but admiration for Darwin, though the men strongly differed in their views of things such as the evolution of the human brain as well as the overall importance of natural selection in driving evolution. Wallace dedicated his classic book *The Malay Archipelago* to Darwin and titled one of his books on evolution *Darwinisn*.

4. *Vestiges* is available from University of Chicago Press, which printed a facsimile in 1994. It's fascinating.

5. The full title of Darwin's work was *On the Origin of Species by Means of Natural Selection or the Preservation of Favoured Races in the Struggle for Life*. The work went through six editions, often with extensive revisions, many of which muddled, rather than clarified. By far the best edition to read is the first edition. Facsimile copies are available. All page numbers that I cite are from a facsimile of the first edition.

6. See http://www.ucmp.berkeley.edu/history/thuxley.html for more on Huxley.

7. Check out the breeds at http://pigeonracing.homestead.com/Pigeon_Breeds.html.

8. An online edition of the essay is found at http://www.ac.wwu.edu/~stephan/malthus/malthus.0.html.

9. See *Origin*, p. 62.

10. J. A. Endler, *Natural Selection in the Wild* (Princeton, NJ: Princeton University Press, 1986).

11. P. T. Boag and P. R. Grant, Intense natural selection in a population of Darwin's finches (*Geospizinae*) in the Galapagos, *Science* 214 (1981): 82–84. The story of the Grants' work on Daphne Major is brilliantly chronicled by Jonathan Weiner in his book *The Beak of the Finch* (New York: Alfred Knopf, 1994).

12. Gregor Mendel did his work on the genetic traits of garden peas about when Darwin published the *Origin*, but Darwin never knew of Mendel's work, which was not really appreciated until the turn of the twentieth century.

13. *Origin*, p. 489.

14. See Kevin Padian, "Darwin's Enduring Legacy," *Nature* 451 (February 7, 2008): 632–34, for a brief overview. For a deeper treatment, see Michael T. Ghiselin, *The Triumph of the Darwinian Method* (Chicago: University of Chicago Press, 1969). Also see Ernst Mayr's excellent overview, *One Long Argument: Charles Darwin and the Genesis of Modern Evolutionary Thought* (Cambridge, MA: Harvard University Press, 1991).

15. This example is described on pp. 73–74 of *Origin*.

16. Ibid., p. 67.

17. Ibid., p. 73, emphasis mine.

18. Abridgements of papers by Möbius and Forbes are found in E. J. Kormondy, *Readings in Ecology* (Englewood Cliffs, NJ: Prentice-Hall, 1965).

19. See Robert A. Croker, *Stephen Forbes and the Rise of American Ecology* (Washington, DC: Smithsonian Institutions Press, 2001).

20. S. A. Forbes, "The Lake as a Microcosm," *Bulletin of the Illinois State Natural History Survey* 15 (1925): 537–50; the quote is on p. 549. This seminal paper was originally published in 1887 by the Peoria Scientific Association in their bulletin.

Chapter 6 The Twentieth Century: Ecology Comes of Age

1. The biological species concept (BSC) has dominated evolutionary biology since the 1940s. It is not the only species definition but it has worked well for most animals and it has strong heuristic value, as it relies on reproductive isolation. It is currently strongly challenged by species concepts based more on analysis of molecular data. Dobjhansky's classic book describing the concept is *Genetics and the Origin of Species* (New York: Columbia University Press, 1937). Mayr's classic book is *Systematics and the Origin of Species* (New York: Columbia University Press, 1942).

2. Simpson's classic book is *Tempo and Mode in Evolution* (New York: Columbia University Press, 1944).

3. See R. E. Croker, *Pioneer Ecologist: The Life and Work of Victor Ernest Shelford* (Washington, DC: Smithsonian Institution Press, 1991).

4. V. E. Shelford, *The Ecology of North America* (Urbana: University of Illinois Press, 1963).

5. See http://www.nature.org/?src=t1 and http://www.nature.org/aboutus/history/.

6. Phytosociology was initially developed in Europe, particularly by the Swiss botanist and ecologist Josias Braun-Blanquet (1884–1980). The Braun-Blanquet methodology was adopted by U.S. plant ecologists.

7. More on Clements is found at http://www.history.ucsb.edu/projects/westcampus/clements/bio.htm. Also see F. E. Clements, *Plant Succession and Indicators* (New York: H. W. Wilson, 1928).

8. More on Gleason is found at http://sciweb.nybg.org/science2/libr/finding_guide/gleapap.asp. Also see H. A. Gleason, "The Individualistic Concept of the Plant Association," *Bulletin of the Torrey Botanical Club* 53 (1926): 1–20.

9. M. Nicholson and R. P. McIntosh, "H. A. Gleason and the Individualistic Hypothesis Revisited," *Bulletin of the Ecological Society of America* 83 (2002): 133–42.

10. See R. T. Brown and J. T. Curtis, "The Upland Conifer-Hardwood Forests of Northern Wisconsin," *Ecological Monographs* 22 (1952): 217–34; and J. R. Bray and J. T. Curtis, "An Ordination of the Upland Forest Communities of Southern Wisconsin," *Ecological Monographs* 27 (1957): 325–49.

11. See R. H. Whittaker, "Gradient Analysis of Vegetation," *Biological Reviews* 42 (1967): 207–64; and R. H. Whittaker and W. A. Niering, "Vegetation

of the Santa Catalina Mountains, Arizona (II): A Gradient Analysis of the South Slope," *Ecology* 46 (1965): 429–52.

12. Elton's *Animal Ecology* has been reprinted and reissued by University of Chicago Press (2001). Elton authored another book, *The Ecology of Invasions of Animals and Plants* (New York: Methuen, 1958), that was far ahead of its time, a tribute to Elton's sweeping insights. Invasive species are a major concern for ecologists today, and Elton's classic work is widely cited. See "The Book That Began Invasion Ecology," a review by A. Ricciardi and H. J. MacIsaac, *Nature* 452 (2008): 34.

13. Elton, *Animal Ecology* (Chicago: University of Chicago Press, 2001), p. 56.

14. Colinvaux's book is available from Princeton University Press.

15. R. L. Lindeman, "The Trophic-Dynamic Aspect of Ecology," *Ecology* 23 (1942): 399–418.

16. H. T. Odum, "Trophic Structure and Productivity of Silver Springs, Florida," *Ecological Monographs* 27 (1957): 55–112.

17. The failure of energy conversions to be 100% efficient is called the law of entropy, or the second law of thermodynamics. Entropy is defined as the tendency for systems to move from order to disorder. Examples abound. Look at your own automobile when you drive it new off the lot and then look at it months and years later. Unless you periodically add energy to clean and repair the car, it "degrades." Such is true of living bodies and of ecosystems. If you don't eat you will die, decompose, and the elegant order and complexity that was apparent in your living corpus will have been gradually degraded and dispersed as heat (feeding numerous organisms as the process occurred). If ecosystems do not receive energy input, they, too, will degrade. Ecosystems are nothing more than collections of living organisms, all of which, without exception, are subject to the laws of thermodynamics.

18. There is an outstanding account of the history of ecosystem ecology in a book by Joel B. Hagen, *An Entangled Bank: The Origins of Ecosystem Ecology* (New Brunswick, NJ: Rutgers University Press, 1992).

19. H. G. Andrewartha and L. C. Birch, *The Distribution and Abundance of Animals* (Chicago: University of Chicago Press, 1954); David Lack, *The Natural Regulation of Animal Numbers* (Oxford: Clarendon Press, 1954).

20. D. Lack, *Darwin's Finches* (1947; repr. Cambridge: Cambridge University Press, 1983). In an interesting note on how science works, Lack did not observe obvious competition or segregation among finches in his first trip to the islands. That was because it was very rainy and food for the birds was abundant. Only on his second trip, when drier conditions prevailed, were the differences among the birds fully apparent.

21. G. E. Hutchinson, "Concluding Remarks," *Cold Spring Harbor Symposia on Quantitative Biology* 22 (1957): 415–27; G. E. Hutchinson, "Homage to Santa Rosalina, or Why Are There So Many Kinds of Animals?" *American Naturalist* 53 (1959): 145–59.

22. The English (really European) wood warbler (*Phylloscopus sibilatrix*) that Gilbert White observed (recall chapter 4) is in the family of old world warblers, Sylviidae. The wood-warblers in North America, the ones discussed in this study, are unrelated to the Old World warblers and are placed in the family Parulidae.

23. R. H. MacArthur, "Population Ecology of Some Warblers of Northeastern Coniferous Forests," *Ecology* 39 (1958): 599–619.

24. A foraging niche is the total range of food resources utilized by a species. It may be broad but is frequently constrained by competition with other species.

25. R. MacArthur and E. O. Wilson, *The Theory of Island Biogeography* (Princeton, NJ: Princeton University Press, 1967).

26. See http://www.lternet.edu/ for information on all current LTER sites and research that is occurring at them.

27. See Daniel B. Botkin's outstanding book *Discordant Harmonies: A New Ecology for the Twenty-first Century* (Oxford: Oxford University Press, 1990). Botkin does a fine job of tracing how ecologists have viewed balance and equilibrium in more detail than I can include in this book.

Chapter 7 A Visit to Bodie: Ecological Space and Time

1. There are numerous websites featuring Bodie. See http://www.bodie.com/, for example. Also available are two engaging books, *A Trip to Bodie Bluff* by J. R. Browne (reprint, Golden, CO: Outbooks, 1981), and *The Ghost Town of Bodie* (Bishop, CA: Chalfont Press, 1967).

2. J. J. Audubon, *The Birds of North America* (Philadelphia: J. B. Chevalier, 1841). Numerous reprints are readily available.

3. Even though I am a firm believer in the individualistic plant community, etc., in my book *A Field Guide to Eastern Forests* (Boston: Houghton Mifflin, 1988) I found it useful to categorize forests in a manner consistent with previous classifications. It is indeed a matter of scale. No two oak-hickory forests are alike but oak-hickory forests are recognizably distinct from beech-maple forests.

4. See D. Worcester, *Nature's Economy: A History of Ecological Ideas* (Cambridge: Cambridge University Press, 1977) for an engaging account of Thoreau's contributions to ecology as well as those of numerous others.

5. E. P. Odum, "The Strategy of Ecosystem Development," *Science* 164 (1969): 262–70. This paper was quite remarkable. In many ways it combined the many years of research on ecological energetics and succession that Odum and his students had done with a good dose of Clementsian teleology. Ecologists have been and still are struggling with the concept of ecosystem stability and how it relates to biodiversity.

6. See F. N. Egerton, "Changing Concepts of the Balance of Nature," *Quarterly Review of Biology* 48 (1973): 322–50.

7. J. E. Lovelock, *Gaia: A New Look at Life on Earth* (Oxford: Oxford University Press, 1979).

8. See H. R. Delcourt and P. A. Delcourt, "Quaternary Landscape Ecology: Relevant Scales of Space and Time," *Landscape Ecology* 2 (1988): 23–44.

9. R. J. G. Savage and M. R. Long, *Mammal Evolution: An Illustrated Guide* (New York: Facts on File, 1986).

10. L. G. Marshall et al., "Mammalian Evolution and the Great American Interchange," *Science* 215 (1982): 1351–57.

11. R. R. Viet and W. R. Petersen, *Birds of Massachusetts* (Lincoln, MA: Massachusetts Audubon Society, 1993).

12. See J. Wu and O. L. Loucks, "From Balance of Nature to Hierarchical Patch Dynamics: A Paradigm Shift in Ecology," *Quarterly Review of Biology* 70 (1995): 439–66. I used a similar title for a lecture and article upon which this chapter is based. But after much thought, I do not accept the contention that a paradigm shift is involved as much as I believe ecologists finally realized what they were looking at and measuring and were thus able to shake off a bias that had been lingering far too long. Also see S.T.A. Pickett and R. S. Ostfeld, "The Shifting Paradigm in Ecology," in R. L. Knight and S. F. Bates, eds., *A New Century for Natural Resource Management* (Washington, DC: Island Press, 1995), pp. 261–79; and S.T.A. Pickett, V. T. Parker, and P. L. Fiedler, "The New Paradigm in Ecology: Implications for Conservation Biology above the Species Level," in P. L. Fiedler and S. K. Jain, eds., *Conservation Biology* (New York: Chapman and Hall, 1992), pp. 65–88.

13. There are numerous examples, and I summarize some in my book *A Neotropical Companion* (Princeton, NJ: Princeton University Press, 1997).

14. See P. L. Marks, "On the Origin of the Field Plants of the Northeastern United States," *American Naturalist* 122 (1983): 210–28, for an excellent discussion of the evolutionary histories of successional plant species.

15. L. White, Jr., "The Historical Roots of Our Ecological Crisis," *Science* 155 (1967): 1203–6.

16. W. Cronin, *Changes in the Land: Indians, Colonists, and the Ecology of New England* (New York: Hill and Wang, 1983).

17. J. E. Mosimann and P. S. Martin, "Simulating Overkill by Paleoindians," *American Scientist* 63 (1975): 304–13.

18. P. Matthiessen, *Wildlife in America* (NY: Viking Penguin Inc., 1959; revised 1987).

19. The Breeding Bird Survey is an effort to annually monitor bird populations breeding within North America. It is coordinated at the USGS Patuxent Wildlife Research Center in Maryland. See http://www.pwrc.usgs.gov/BBS/ for details.

20. J. J. Audubon, *The Birds of North America*, vol. 2 (Philadelphia: J. B. Chevalier, 1841), p. 34.

21. The conversion of New England forest to pasture and agriculture is wonderfully illustrated in *New England Forests through Time* by D. R. Foster and J. F. O'Keefe (Cambridge, MA: Harvard University Press, 2000). The illustrations are from the magnificent dioramas at the Harvard Forest in Petersham, MA.

22. E. H. Bucher, "The Causes of the Extinction of the Passenger Pigeon," in D. M. Power, ed., *Current Ornithology*, vol. 9 (New York: Plenum Press, 1992), 1–36.

23. Viet and Petersen, *Birds of Massachusetts* (see n. 11 above).

24. Matthiessen, *Wildlife in America* (see n. 18 above).

CHAPTER 8 ECOLOGY AND EVOLUTION: PROCESS AND PARADIGM

1. This book is long out of print. It was published in 1965 by Yale University Press and is still available in some libraries.

2. These include such distinguished ecologists as P. Klopfer, L. Slobodkin, F. E. Smith, E. S. Deevey, R. H. MacArthur, A. J. Kohn, W. T. Edmondson, H. T. Odum, J. L. Brooks, and I. M. Newell.

3. E. P. Odum, *Fundamentals of Ecology* (Philadelphia: W. B. Saunders, 1953). This text moved ecology into the mainstream of biology offerings in numerous institutions.

4. E. P. Odum and G. W. Barrett, *Fundamentals of Ecology*, 5th ed. (Belmont, CA: Thomson Brooks/Cole, 2005).

5. W. C. Allee et al., *Principles of Animal Ecology* (Philadelphia: W. B. Saunders Co.) p. 695.

6. Ibid., p. 729.

7. J. Tewksbury and G. Nabhan, "Directed Deterrence by Capsaicin in Chilies," *Nature* 412 (July 26, 2001): 403.

8. Mayr discusses how- and why-type questions in several of his books. There is an extensive treatment in *The Growth of Biological Thought* (Cambridge: Belknap Press, 1982). Among other books is *Toward a New Philosophy of Biology: Observations of an Evolutionist* (Cambridge, MA: Belknap Press, 1988); see especially essay 17. Also see *This Is Biology: The Science of the Living World* (Cambridge, MA: Belknap Press, 1997).

9. The deep ocean hot vents are not dependent on photosynthesis, as they are ecosystems that thrive with heat as the base of energy. Bacteria use heat to accomplish chemosynthesis. But deep ocean hot vents are very limited in distribution and represent a miniscule amount of life support compared with global photosynthesis.

10. R. P. Balda and A. C. Kamil, "A Comparative Study of Cache Recovery by Three Corvid Species," *Animal Behavior* 38 (1989): 486–95.

11. N. J. Emery and N. S. Clayton, "The Mentality of Crows: Convergent Evolution of Intelligence in Corvids and Apes," *Science* 306 (2004): 1903–7.

12. P. D. Smallwood and W. D. Peters, "Grey Squirrel Food Preferences: The Effects of Tannin and Fat Concentration," *Ecology* 67 (1986): 168–74. See also M. Steele and P. Smallwood, "What Are Squirrels Hiding?" *Natural History*, October 1994, pp 40–45.

13. S. J. Gould, "Evolution and the Triumph of Homology, or Why History Matters," *American Scientist* 74 (1986): 60–69.

14. E. H. Bucher, "The Causes of the Extinction of the Passenger Pigeon," (see chapter 7, n. 22 above).

15. W. J. McShea, "The Influence of Acorn Crops on Annual Variation in Rodent and Bird Populations," *Ecology* 81 (2000): 228–38.

16. See http://www.fs.fed.us/ne/delaware/atlas/web_atlaso.html as an example of how various models predict changes in tree species distribution with climate change scenarios.

17. C. G. Jones et al., "Chain Reactions Linking Acorns to Gypsy Moth Outbreaks and Lyme Disease," *Science* 279 (1988): 1023–26.

18. J. S. Elkinton et al., "Interactions Among Gypsy Moths, White-Footed Mice, and Acorns," *Ecology* 77 (1996): 2332–42.

CHAPTER 9 BE GLAD TO BE AN EARTHLING

1. R. Dawkins, *The Selfish Gene* (Oxford: Oxford University Press, 1976). It is difficult to overstate the impact of this book on making evolutionary biology comprehensible to educated general readers as well as influencing how biologists perceived natural selection. I strongly recommend it as well as Dawkins's *The Blind Watchmaker* (New York: W.W. Norton, 1986). In addition, the interested reader would be well advised to read the now classic volume by George C. Williams, *Adaptation and Natural Selection: A Critique of Some Current Evolutionary Thought* (Princeton, NJ: Princeton University Press, 1966). Natural selection is more tricky to comprehend than it seems. These books make it extremely clear.

2. I am very well aware that humans cooperate, form deep friendships, love, commit to relationships, embrace, and are capable of extreme generosity. I like to think I fit into such a description of the positive aspects of my species. But the plasticity of human behavior and the capacity for love and commitment are at least in part explained by areas of natural selection theory such as kin selection, group selection, and reciprocal altruism. Knowing this does not reduce the value of those traits but merely illuminates their origins.

3. Fred Hoyle (1915–2001) was an iconoclast who had a remarkable career. He is widely known for his opposition to the concept of an expanding universe, and indeed, it was he would coined the term "big bang," meant to be a cynical description that turned into a widely used term, when evidence accumulated that the universe is really expanding. He was an opponent of natural selection, in large part due to his own work on the improbability of the carbon atom having the exact properties that it has. He argued (quite unsuccessfully) that the fossil dinosaur/bird *Archaeopteryx* is a fraud. He believed that life colonized Earth from space (rather than evolved on Earth), and that diseases such as syphilis and AIDS came to Earth through seeding by a passing comet. Hoyle also wrote some engaging science fiction, the best of which, in my opinion, was *The Black Cloud* (1957). Hoyle's autobiography, *Home Is Where the Wind Blows: Chapters from a Cosmologist's Life*, was published in 1994. I also recommend Simon Mitton's

Conflict in the Cosmos: Fred Hoyle's Life in Science (Washington, DC: Joseph Henry Press, 2005).

4. The anthropic principle is discussed at length in numerous websites and books. See, for example, http://ourworld.compuserve.com/homepages/rossuk/c-anthro. htm for a good overview with numerous links to books and other websites.

5. The evidence for water having flowed on Mars is abundantly discussed in journals such as *Science* and *Nature* as well as popular magazines such as *Astronomy* and *Sky and Telescope*. There are also websites, including those supported by NASA, that provide explanations and photographs.

6. There are numerous websites, including NASA's, that contain striking images of Jupiter and its major moons.

7. It is sobering to read Martin Rees's book *Our Final Hour* (New York: Basic Books, 2003). Rees presents a series of clearly reasoned arguments suggesting that humanity might well not survive through the twenty-first century.

8. Dozens of websites describe the Drake equation. See, for example, http://www.activemind.com/Mysterious/Topics/SETI/drake_equation.html.

9. P. D. Ward and D. Brownlee, *Rare Earth: Why Complex Life Is Uncommon in the Universe* (New York: Copernicus, 2000). This book is extremely fun to read and deeply insightful.

10. There is an excellent discussion of the obliquity issue in Ward and Brownlee, *Rare Earth*, pp. 222–26 (n. 9 above).

11. See, for example, http://www.psi.edu/projects/moon/moon.html for an overview with illustrations. See also P. D. Spudis, "The Moon," in J. K. Beatty, C. C. Petersen, and A. Chaikin, eds., *The New Solar System* (Cambridge, MA: Sky Publishing, 1999); R. M. Canup and K. Righter, *Origin of the Earth and Moon* (Tucson: University of Arizona Press, 2000); and J. Melosh, "A New Model Moon," *Nature* 412 (2001): 694–95.

12. See http://seds.org/archive/sl9/sl9.html for images of the impact.

13. See http://www.astronomycafe.net/qadir/ask/a11789.html for further explanation and illustration.

14. The theoretical physicist and cosmologist Lee Smolin suggested how a process such as natural selection could be at work within a multiverse of universes in his engaging book *The Life of the Cosmos* (Oxford: Oxford University Press, 1997).

CHAPTER 10 LIFE PLAYS THE LOTTERY

1. My lectures, including one upon which this chapter is partly based, can be heard as "Behold the Mighty Dinosaur," published by Modern Scholar. See http://www.rbfilm.com/index.cfm?fuseaction=scholar.show_course&course_id=101.

2. This reality of time and evolution was conveniently ignored in one of the finest early depictions of dinosaurs on the big screen. In the 1940 Disney animated

film *Fantasia,* a dramatic sequence included a very ferocious *Tyrannosaurus* pursuing and killing a lumbering *Stegosaurus,* as rain poured down. Wow, it was good. Not accurate, but wonderful nonetheless.

3. The Ordovician, Devonian, and Permian events occurred in the Paleozoic era and the other two were in the Mesozoic era. The greatest extinction of all time was the Permian event. See http://www.space.com/scienceastronomy/planetearth/extinction_sidebar_000907.html for a brief summary of each.

4. The term "non-avian" dinosaur is now in wide use, as most evolutionary biologists agree that birds evolved directly from one branch of carnivorous (theropod) dinosaurs. Non-avian dinosaur refers to those dinosaurs that were not birds. Non-avian is used because when applying the modern method of classification, called phylogenetic systematics or simply "cladistics," birds are firmly nested (no pun intended) within dinosaurs and thus birds are a form of dinosaur. Some birds but no non-avian dinosaurs survived the Cretaceous extinction event. For more information consult *Glorified Dinosaurs* by Luis M. Chiappe (Hoboken, NJ: Wiley, 2007).

5. At the beginning of the Mesozoic era, 248 million years ago, not only had Earth just suffered its worst mass extinction event but all continents were fused comprising a supercontinent named Pangaea. During the course of the Mesozoic era and continuing today, this continent broke apart, first as Laurasia to the north and Gondwana to the south. These subsequently divided into today's distribution of continents, all of which continue to move in various directions as they ride atop tectonic plates.

6. There are dozens of websites that illustrate dinosaurs as well as many popular books. A good general book, profusely illustrated, is *Dinosaurs* by T. R. Holtz, illustrated by L. V. Rey (New York: Random House, 2007).

7. The modern horse, *Equus,* is the last remnant of a once flourishing diverse array of species that included both browsers and grazers. As grasses increased, grazing species did too. See http://chem.tufts.edu/science/evolution/HorseEvolution.htm for examples.

8. G. Poinar, Jr., and Roberta Poinar, *What Bugged the Dinosaurs? Insects, Disease and Death in the Cretaceous* (Princeton, NJ: Princeton University Press, 2007).

9. Good general references that discuss both the volcanism and impact hypotheses include J. D. Archibald, *Dinosaur Extinction and the End of an Era: What the Fossils Say* (New York: Columbia University Press, 1996); and D. E. Fastovsky and D. B. Weishampel, *The Evolution and Extinction of the Dinosaurs* (2nd ed., Cambridge: Cambridge University Press, 2005). For a first-hand account of the impact hypothesis, see W. Alvarez, *T. rex and the Crater of Doom* (Princeton, NJ: Princeton University Press, 1997).

10. D. S. Robertson et al., "Survival in the First Hours of the Cenozoic," *Geological Society of America Bulletin* 116 (2004): 760–68. See also D. E. Fastovsky,

"The Extinction of the Dinosaurs in North America," *GSA Today* 15 (2005): 4–10.

11. S. Sahney and M. J. Benton, "Recovery from the Most Profound Extinction of All Time," *Proceedings of the Royal Society B* 275, no. 1636 (2008).

12. S. J. Gould, *Wonderful Life: The Burgess Shale and the Nature of History* (New York: W.W. Norton, 1989). This book was a best seller but became controversial when one of the persons Gould discussed at length, Simon Conway Morris, published his own account of the Burgess Shale fauna with a radically different interpretation from Gould's. Morris's book is *The Crucible of Creation: The Burgess Shale and the Rise of Animals* (Oxford: Oxford University Press, 1998). Morris disagrees not only with Gould's interpretation of the Burgess Shale but with contingency's importance, as described by Gould.

13. There are excellent illustrations of the "dinosauroid" in the final chapter of Dale Russell's engaging book *An Odyssey in Time: The Dinosaurs of North America* (Minocqua, WI: NorthWord Press, 1989).

14. It is useful to consult Ernst Mayr's splendid book *The Growth of Biological Thought* (Cambridge, MA: Belknap Press, 1982) for an enlightening discussion of the *scala naturae* and its influence in the emergence of evolutionary thought.

15. D. M. Raup, *Extinction: Bad Genes or Bad Luck?* (New York: W.W. Norton, 1991). For a somewhat more technical discussion, see D. M. Raup, "Biological Extinction in Earth History," *Science* 231 (1986): 1528–33.

16. R. Stone, "Preparing for Doomsday," *Science* 319 (2008): 1326–29.

CHAPTER 11 WHY GLOBAL CLIMATE IS LIKE NEW ENGLAND WEATHER

1. See, for example, M. G. Barbour and W. D. Billings, eds., *North American Terrestrial Vegetation* (Cambridge: Cambridge University Press, 1988).

2. See http://cpluhna.nau.edu/Biota/merriam.htm for more.

3. P. D. Ward, *Out of Thin Air: Dinosaurs, Birds, and Earth's Ancient Atmosphere* (Washington, DC: Joseph Henry Press, 2006).

4. Industrial melanism is the term describing the natural selection of dark-colored moths (mostly of *Biston bitularia*) over light morphs (the prevalent phenotype) when tree bark got covered with soot from coal burning. It is described in virtually all evolution texts and on numerous websites.

5. See http://cdiac.ornl.gov/trends/co2/contents.htm.

6. See http://www.nasa.gov/centers/ames/news/releases/2002/02_60AR.html.

7. C. Rosenzweig et al., "Attributing Physical and Biological Impacts to Anthropogenic Climate Change," *Nature* 453 (2008): 353–57.

8. T. Root, "Energy Constraints on Avian Distributions and Abundances," *Ecology* 69 (1988): 330–39.

9. http://www.abcbirds.org/conservationissues/globalwarming/.

Chapter 12 Taking It from the Top—or the Bottom

1. To explore this notion in greater detail, see R. M. May, S. A. Levin, and George Sugihara, "Ecology for Bankers," *Nature* 451 (2008): 893–95.

2. B. R. Silliman and M. D. Bertness, "A Trophic Cascade Regulates Salt Marsh Primary Productivity," *Proceedings of the National Academy of Science* 99 (2002): 10500–10505.

3. N. G. Hairston, F. E. Smith, and L. B. Slobodkin, "Community Structure, Population Control, and Competition," *American Naturalist* 94 (1960): 421–25.

4. P. R. Ehrlich and L. C. Birch, "The Balance of Nature and Population Control," *American Naturalist* 101 (1967): 97–107.

5. J. A. Estes et al., "Killer Whale Predation on Sea Otters Linking Oceanic and Nearshore Ecosystems," *Science* 282 (1998): 473–76.

6. J. Terborgh, "Keystone Plant Resources in the Tropical Forest," in M. E. Soule, ed., *Conservation Biology: The Science of Scarcity and Diversity* (Sunderland, MA: Sinauer, 1986).

7. See W. Bond, "Keystone Species—Hunting the Snark?" *Science* 292 (2001): 63–64.

8. E. L. Braun, *Deciduous Forests of Eastern North America* (New York: Hafner, 1972).

9. R. T. Paine, "Food Web Complexity and Species Diversity," *American Naturalist* 100 (1966): 65–75.

10. R. T. Paine, "Food Webs: Linkage, Interaction Strength and Community Infrastructure," *Journal of Animal Ecology* 49 (1980): 667–85.

11. P. P. Marquis and C. J. Whelan, "Insectivorous Birds Increase Growth of White Oak through Consumption of Leaf-chewing Insects," *Ecology* 75 (1994): 2007–14.

12. M. Murakami and S. Nakano, "Species-specific Bird Functions in a Forest-Canopy Food Web," *Proceedings of the Royal Society of London B* 267 (2000): 1597–1600.

13. M. B. Kalka, A. R. Smith, and E. K. V. Kalko, "Bats Limit Arthropods and Herbivory in a Tropical Forest," *Science* 320 (2008): 71.

14. J. Terborgh et al., "Ecological Meltdown in Predator-free Forest Fragments," *Science* 294 (2001): 1923–26.

15. K. A. M. Engelhardt and M. E. Ritchie, "Effects of Macrophyte Species Richness on Wetland Ecosystem Functioning and Services," *Nature* 411 (2001): 687–89.

16. D. A. Levin, "Alkaloid-bearing Plants: An Ecogeographic Perspective," *American Naturalist* 110 (1976): 261–84.

17. R. S. Ostfeld and R. D. Holt, "Are Predators Good for Your Health? Evaluating Evidence for Top-down Regulation of Zoonotic Disease Reservoirs," *Frontiers of Ecology and the Environment* 2 (2004): 13–20.

18. S. L. Pimm, *Food Webs* (London: Chapman and Hall, 1982).

19. See http://www.innovations-report.com/html/reports/life_sciences/report 31960.html.

CHAPTER 13 FOR THE LOVE OF BIODIVERSITY (AND STABLE ECOSYSTEMS?)

1. See http://dwb.adn.com/front/story/4110831p-4127072c.html for the story. There is also a book and film about the incident.

2. See http://www.fws.gov/laws/lawsdigest/ESACT.HTML for details.

3. See S. J. O'Brien et al., "The Cheetah Is Depauperate in Genetic Variation," *Science* 221 (1983): 459–62; and S. J. O'Brien et al., "Genetic Basis for Species Vulnerability in the Cheetah," *Science* 227 (1985): 1428–34. But also see Roger Lewin's article, "A Strategy for Survival?" in *New Scientist* (issue 2017, February 17, 1996) for an account of how acrimonious the cheetah conservation debate became.

4. T. L. Erwin, "Tropical Forests: Their Richness in Coleoptera and Other Arthropod Species," *Coleopterists' Bulletin* 36 (1982): 74–75; T. L. Erwin, "The Tropical Forest Canopy: The Heart of Biotic Diversity," in E. O. Wilson, ed., *Biodiversity* (Washington, DC: National Academy Press, 1988).

5. P. R. Ehrlich and E. O. Wilson, "Biodiversity Studies: Science and Policy," *Science* 253 (1991): 758–62.

6. R. M. May, "How Many Species Are There on Earth?" *Science* 241 (1988): 1441–49.

7. B. Groombridge and M. D. Jenkins, *World Atlas of Biodiversity* (Berkeley: University of California Press, 2002).

8. Ibid.

9. B. W. Brook, N. S. Sodhi, and P. K. L. Ng, "Catastrophic Extinctions Follow Deforestation in Singapore," *Nature* 424 (2003): 420–23.

10. See http://www.fort.usgs.gov/resources/education/bts/impacts/birds.asp# gotohere1 for a thorough review of the full impact of the brown treesnake as it affected birds.

11. P. D. Walsh et al., "Catastrophic Ape Decline in Western Equatorial Africa," *Nature* 422 (2003): 611–14.

12. A. Balmford et al., "Conservation Conflicts across Africa," *Science* 291 (2001): 2616–19.

13. E. O. Wilson, *The Diversity of Life* (Cambridge, MA: Belknap Press, 1992).

14. J. C. Kricher, "Bird Species Diversity: The Effect of Species Richness and Equitability on the Diversity Index," *Ecology* 53 (1972): 278–82.

15. The results of the conference are summarized in M. Loreau et al., "Biodiversity and Ecosystem Functioning: Current Knowledge and Future Challenges," *Science* 294 (2001): 804–8; and M. Loreau, S. Naeem, and P. Inchausti, *Biodiversity and Ecosystem Functioning: Synthesis and Perspectives* (Oxford: Oxford University Press, 2002).

16. These depictions were summarized in F. Schläpfer and B. Schmid, "Ecosystem Effects of Biodiversity: A Classification of Hypotheses and Cross-system Exploration of Empirical Results," *Ecological Applications* 9 (1999): 893–912.

17. For example, A. P. Kinzig, S. W. Pacala, and D. Tilman, *The Functional Consequences of Biodiversity* (Princeton, NJ: Princeton University Press, 2002).

18. P. R. Ehrlich and A. Ehrlich, *Extinction: The Causes and Consequences of the Disappearance of Species* (New York: Random House, 1981); E. Marris, "What to Let Go," *Nature* 450 (2008): 152–55.

19. Many papers deal with this point. Here's some of the most insightful: S. Naeem, "Ecosystem Consequences of Biodiversity Loss: The Evolution of a Paradigm," *Ecology* 83 (2002): 1537–52; S. Naeem et al., "Declining Biodiversity Can Alter the Performance of Ecosystems," *Nature* 368 (1994): 734–36; S. Naeem et al., "Empirical Evidence That Declining Species Diversity May Alter the Performance of Terrestrial Ecosystems," *Transactions of the Royal Society of London B* 347 (1995): 249–62. S. Naeem et al., "Biodiversity and Ecosystem Functioning: Maintaining Natural Life Support Processes," *Issues in Ecology* 4 (Washington, DC: Ecological Society of America, 1999); D. Tilman and J. A. Downing, "Biodiversity and Stability in Grasslands," *Nature* 367 (1994): 363–65; D. Tilman, D. Wedin, and J. Knops, "Productivity and Sustainability Influenced by Biodiversity in Grassland Ecosystems," *Nature* 379 (1996): 718–20; D. Tilman et al., "Diversity and Productivity in a Long-term Grassland Experiment," *Science* 294 (2001): 843–45.

20. B. J. Cardinale, M. A. Palmer, and S. L. Collins, "Species Diversity Enhances Ecosystem Functioning through Interspecific Facilitation," *Nature* 415 (2002): 426–29.

21. Tilman et al., "Diversity and Productivity in a Long-term Grassland Experiment," (see n. 19 above).

22. R. M. May, "Will a Large Complex System Be Stable?" *Nature* 238 (1972): 413–14; R. M. May, *Stability and Complexity in Model Ecosystems* (Princeton, NJ: Princeton University Press, 1973).

23. K. S. McCann, "The Diversity-Stability Debate," *Nature* 405 (2000): 228–33.

24. S. L. Pimm, *Food Webs* (London: Chapman and Hall, 1982); S. L. Pimm, *The Balance of Nature* (Chicago: University of Chicago Press, 1991).

25. T. A. Kennedy et al., "Biodiversity as a Barrier to Ecological Invasion," *Nature* 417 (2002): 636–38.

26. G. B. De Deyn et al., "Soil Invertebrate Fauna Enhances Grassland Succession and Diversity," *Nature* 422 (2003): 711–13.

27. B. Worm et al., "Impacts of Biodiversity Loss on Ocean Ecosystem Services," *Science* 314 (2006): 787–90.

28. S. Naeem, "Ecosystem Consequences of Biodiversity Loss" (see n. 19 above).

29. L. Gamfeldt, H. Heillebrand, and P. R. Jonsson, "Multiple Functions Increase the Importance of Biodiversity for Overall Ecosystem Functioning," *Ecology* 89 (2008): 1223–31.

CHAPTER 14 FACING MARLEY'S GHOST

1. G. C. Daily, *Nature's Services* (Washington, DC: Island Press, 1967). Also see G. C. Daily et al., "Ecosystem Services: Benefits Supplied to Human Societies by Natural Ecosystems," in *Issues in Ecology*, no. 2 (Washington, DC: Ecological Society of America, 1997).

2. P. Sears, "Ecology—a Subversive Subject," *Bioscience* 14 (1964): 11.

3. For a quick summary of Locke's views, see http://www.philosophypages.com/hy/4n.htm.

4. L. White, Jr., "The Historical Roots of Our Ecological Crisis," *Science* 155 (1967): 1203–7.

5. G. Hardin, "The Tragedy of the Commons," *Science* 162 (1968): 1243–48.

6. P. Harrison and F. Pearce, *AAAS Atlas of Population and Environment* (Berkeley: University of California Press, 2000).

7. E. O. Wilson, *The Future of Life* (New York: Knopf, 2002).

8. W. E. Rees and M. Wackernagel, "The Ecological Footprint," in A. M. Jansson et al., eds., *Investing in Natural Capital: The Ecological Economics Approach to Sustainability* (Washington, DC: Island Press, 1994).

9. Wilson, *The Future of Life* (see n. 7 above).

10. P. M. Vitousek et al., "Human Domination of Earth's Ecosystems," *Science* 277 (1997): 494–99.

11. J. N. Galloway et al., "Transformation of the Nitrogen Cycle: Recent Trends, Questions, and Potential Solutions," *Science* 320 (2008): 889–97. The authors of this study suggest that human activities are releasing an inordinate amount of nitrogen in air, water, and land, leading to numerous environmental and human health problems, and they urge an "integrated interdisciplinary approach" to decrease nitrogen-containing waste.

12. Birdlife International, *Threatened Birds of the World* (Barcelona: Lynx Edicions, 2000).

13. S. R. Palumbi, "Humans as the World's Greatest Evolutionary Force," *Science* 293 (2001): 1786–90.

14. Vitousek et al., "Human Domination of Earth's Ecosystems" (see n. 10 above).

15. P. H. Raven, "Science, Sustainability, and the Human Prospect," *Science* 297 (2002): 954–58.

16. M. Milinski, D. Semmann, and H.-J. Krambeck, "Reputation Helps Solve the 'Tragedy of the Commons,'" *Nature* 415 (2002): 424–26.

17. E. Ostrom et al., "Revisiting the Commons: Local Lessons, Global Challenges," *Science* 284 (1999): 278–82.

18. R. Costanza et al., "The Economic Value of the World's Ecosystems," *Nature* 387 (1997): 253–60.

19. A. Balmford et al., "Economic Reasons for Conserving Wild Nature," *Science* 297 (2002): 950–53.

20. Ibid.

21. A. Leopold, *A Sand County Almanac* (Oxford: Oxford University Press, 1949). Available in numerous editions, this wonderful book contains the essay on the land ethic.

22. J. Muir, *The Yosemite* (New York: Century Company, 1912).

23. R. Carson, *Silent Spring* (Boston: Houghton Mifflin Company, 1962).

24. See, for example, P. Singer, *The Expanding Circle: The Ethics of Sociobiology* (New York: Farrar, Strauss, & Giroux, 1981); H. Rolston III, *Philosophy Gone Wild: Essays in Environmental Ethics* (Buffalo, NY: Prometheus, 1986); B. G. Norton, *Why Preserve Natural Variety?* (Princeton, NJ: Princeton University Press, 1987); S. R. Kellert and E. O. Wilson, eds., *The Biophilia Hypothesis* (Washington, DC: Island Press, 1993); and N. S. Cooper and R.C.J. Carling, eds., *Ecologists and Ethical Judgments* (Cambridge: Cambridge University Press, 1996).

25. Norton, *Why Preserve Natural Variety?* (see n. 24 above), p. 135.

26. E. O. Wilson, *Biophilia* (Cambridge, MA: Harvard University Press, 1984).

27. D. H. Janzen, "Tropical Ecological and Biocultural Restoration," *Science* 239 (1988): 243–44.

28. *A Christmas Carol* by Charles Dickens was published on December 19, 1843. Its full title was *A Christmas Carol in Prose, Being a Ghost Story of Christmas*. Numerous editions are available, and it is available also on the web at several sites.

29. Available at http://arts.cuhk.edu.hk/Philosophy/Kant/cpr/.

Epilogue

1. J. C. Kricher, "Nothing Endures but Change: Ecology's Newly Emerging Paradigm," *Northeastern Naturalist* 5 (1998): 165–74.

❧ Index